SOLVING
HOME
PLUMBING
PROBLEMS

SOLVING
HOME
PLUMBING
PROBLEMS

GARY BRANSON

FIREFLY BOOKS

A FIREFLY BOOK

Published by Firefly Books (U.S.) Inc. 2004

First Printing

Publisher Cataloging-in-Publication Data (U.S.)

Branson, Gary D.
    Solving home plumbing problems / Gary Branson.—1st ed.
[144] p. : col. ill. ;   cm.
Includes index.
Summary: Guide to repairing the most common plumbing problems for homeowners, including the basics of plumbing tools and systems, working with pipes,
fixing leaks, faucets and drains, remodeling, and dealing with minor emergencies.
ISBN 1-55297-875-3
ISBN 1-55297-876-1 (pbk.)
1. Plumbing -- Amateurs' manuals. 2. Dwellings -- Maintenance and repair --
Amateurs' manuals. I. Title.
696/.1 21   TH6124.B73   2004

Published in the United States in 2004 by
Firefly Books (U.S.) Inc.
P.O. Box 1338, Ellicott Station
Buffalo, New York, USA
14205

Published in Canada in 2004 by Key Porter Books Limited.

Design: Peter Maher
Electronic formatting: Jean Peters

Printed and bound in Canada

CONTENTS

Introduction

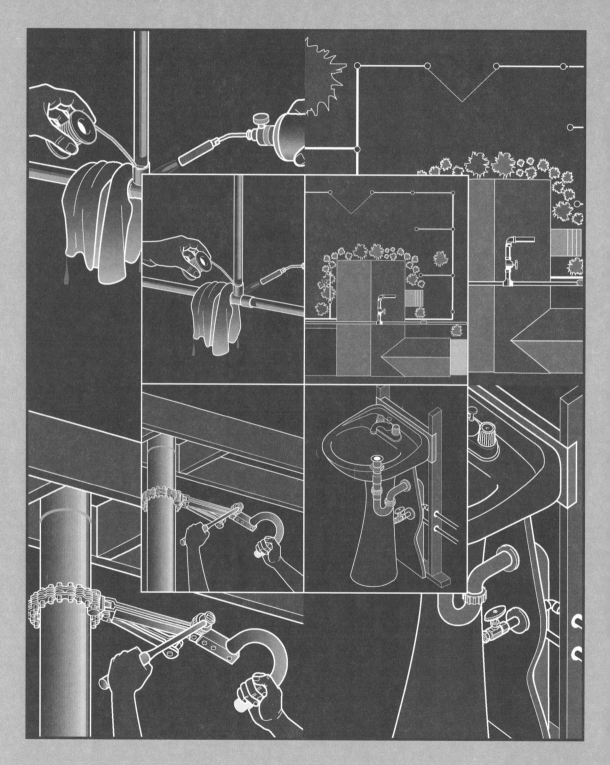

From New York to London, from Los Angeles to Tokyo, the cities of the world share a common beginning: they grew along the shores of major waterways, and for good reason: people need water for survival. Today, modern residential plumbing assumes a daily water usage of 75 to 100 gallons per person, per day. We also need water for commerce, for industry, for agriculture and for transportation. We require a delivery system to deliver water to us, and we need a drain/waste system to remove wastewater.

Among ancient cities, Rome had one of the most sophisticated water-distribution systems. The Romans developed aqueducts to bring potable water—water fit for human consumption—into the city. The word aqueduct comes from "aqua," the Latin word for water, and "duct," the Latin word for a canal or tube.

Lacking the resources and technology to make copper or steel pipe, the Romans made water distribution pipes of lead. Today, health considerations prohibit the use of lead water pipes. There is a modern reminder of the use of lead for pipes: the Latin word for lead is "plumbum," the root of the English word "plumbing."

In this book we help the reader to understand the entire home plumbing system. Because plumbing is one of the most expensive systems in a house, do-it-yourself plumbing repairs and maintenance can result in money in your pocket. This is because plumbers charge a lot for their services. Seventy-five percent of the average repair bill is for labor, 25 percent for materials. The high labor costs translate into significant savings if you do the work yourself. We also emphasize the fact that the greatest savings can result from simple preventive maintenance procedures, which the homeowner can employ both to save money and to avoid unnecessary repairs.

Before you undertake plumbing tasks, it's helpful to know that you can make repairs and replace fixtures without obtaining a permit from the building department. However, for major alterations or additions, such as adding a bathroom, you must obtain a permit and call for periodic inspections. In the U.S., each state has a plumbing code dictated by the state health department intended to protect public health concerns.

Working Safely

Note also two safety tips. When soldering copper pipes, you will be working with a soldering torch with an open flame. Be extremely careful when using an open flame near wood framing or other combustible materials. Keep a water hose or fire extinguisher handy for use if needed. Better still, opt for plastic pipe and fittings to avoid having to use an open flame.

Your house plumbing begins with metal delivery pipes buried in the earth, and exits with a metal pipe waste system. This means that the metal plumbing pipes provide the ultimate electrical ground. To avoid a dangerous or deadly electrical shock, you should not use corded electrical tools on or near metal plumbing pipes. Instead, use battery-powered hand tools. If your home is plumbed with plastic pipe, this warning does not apply, because plastic is not an electrical conductor.

Finally, service people state that more than one-half—one source estimates 75 percent—of appliance service calls come from people who simply forgot to plug an appliance in, or to turn the water and power on. So, before you call for help, make sure you've done these things.

PART I

PLUMBING BASICS

CHAPTER 1

Plumbing Tools

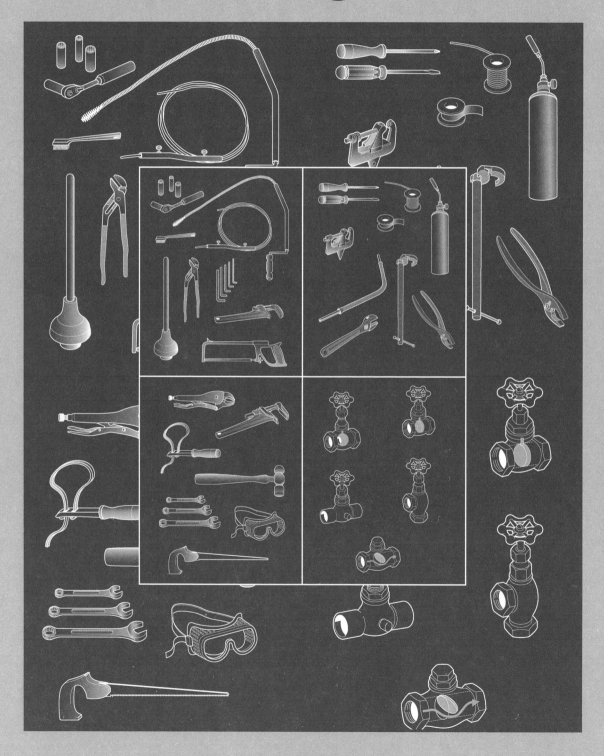

There is an adage in the construction trades that states "By the tools you will know the workman." Having the proper tools is the key to doing any job with workman-like results. Too often, the handyperson will attempt a plumbing project using whatever tools he or she has at hand. The result is that each step of the job becomes doubly difficult and frustrating, and the job results prove less than acceptable.

For example, you may be able to loosen or remove a fitting with a pair of ordinary pliers rather than a wrench, but the pliers are likely to slip on the fitting and damage it—or your knuckles. You can cut copper pipe with a hacksaw, but a tubing cutter will make the job easier. The cut will be straight and clean, ensuring a good fit and a watertight joint.

Here is a true story that demonstrates the bad results of using the wrong tool. A man rented a house when he moved to another state. He came home one night to find the landlord's son had decided to do some soldering on a water pipe. Lacking a soldering torch, the boy had lit a blowtorch and tried to heat the pipe with it. (A blowtorch has a spreader tip, designed to spread the heat over a wide area, rather than the concentrated flame tip of a soldering torch.) The result was that the pipe never got hot: the solder was hanging in clumps from the pipe joint. The wide flame had ignited a nearby floor joist, which was smoking. The landlord's son sat in the basement for hours to ensure the flame had gone out and the house would not burn down. The next day he called a plumber.

THE PLUMBING TOOLBOX

The tools you should assemble in your plumbing toolbox will depend on what sort of plumbing projects you plan to undertake. If you intend to limit your efforts to simple repair jobs such as faucet repair and clearing minor plugs in the drainpipes, you will need a few inexpensive tools.

To do simple jobs, you will need a flanged plunger, or "plumber's friend," a socket wrench with deep sockets, a pipe wrench, water pump (channel-type) pliers, needle-nose pliers, vise-grip pliers, a hacksaw, a wire brush, hex (Allen) wrenches, slot and Phillips screwdrivers and a drain auger. (The tools listed are shown in the "Plumbing Tools" Illustration on page 12)

For more ambitious plumbing projects, you will need a basin wrench (a wrench with a long handle for reaching behind the sink basin to remove a faucet), a tubing cutter, and a soldering torch. Materials you will need are solder, flux, Teflon plumber's tape and a tube of Teflon joint compound. If you have compression-type faucets (stem and seat) you will need a box of assorted washers and O-rings.

It is impossible to stock

Tip Box

Good tools can last forever if you take care of them. Low-end, handheld power tools lack the power to do the job and the quality to last. Some high-end tools, such as half-inch drills, can last up to 40 years. So it's worth the investment even if they cost twice as much as the basic model. Quality doesn't cost, it pays.

Don't store tools in a jumble or junk drawer where they can get jostled and become damaged. Install pegboard in your shop and hang tools from pegs. This will prevent them from being damaged and keep them at hand when a project looms. If you store steel tools in a toolbox, buy some silica gel to absorb moisture and prevent the tools from rusting. Lubricate the moving parts of metal tools, such as the jaws and adjustments on wrenches and pliers. Wipe metal tools with a cloth and fine lubricating oil to prevent rust.

Plumbing Tools

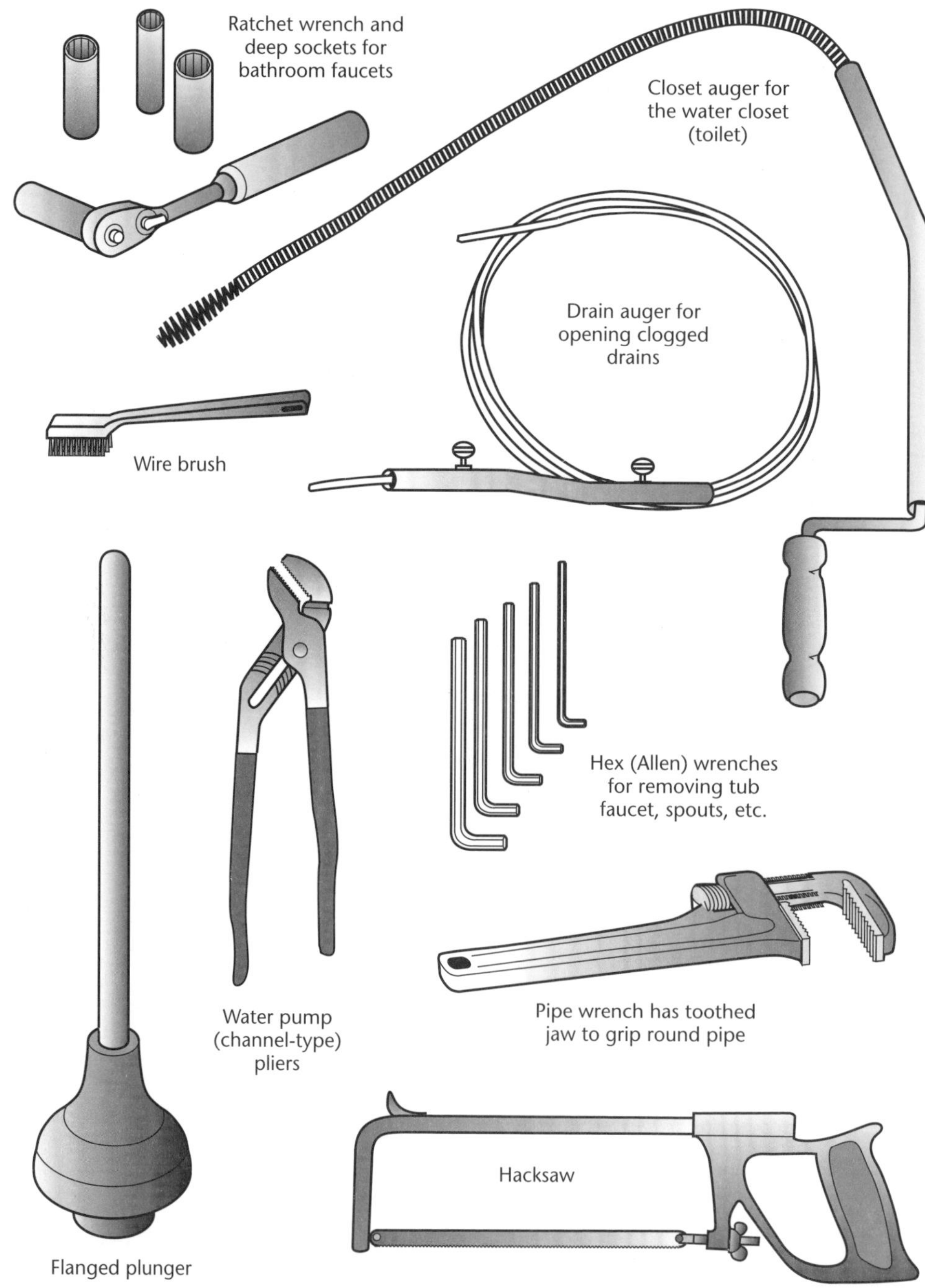

Ratchet wrench and deep sockets for bathroom faucets

Closet auger for the water closet (toilet)

Drain auger for opening clogged drains

Wire brush

Hex (Allen) wrenches for removing tub faucet, spouts, etc.

Water pump (channel-type) pliers

Pipe wrench has toothed jaw to grip round pipe

Hacksaw

Flanged plunger

Plumbing Tools

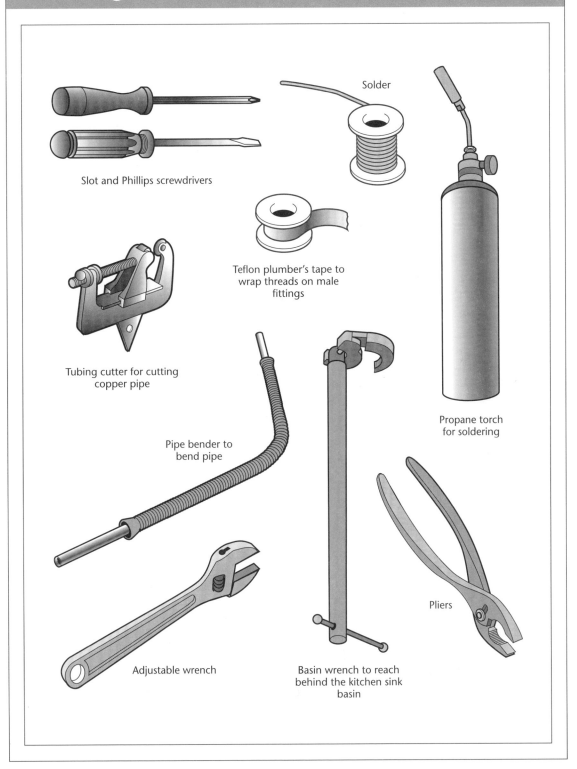

Slot and Phillips screwdrivers

Solder

Teflon plumber's tape to wrap threads on male fittings

Tubing cutter for cutting copper pipe

Pipe bender to bend pipe

Propane torch for soldering

Pliers

Adjustable wrench

Basin wrench to reach behind the kitchen sink basin

Plumbing Tools

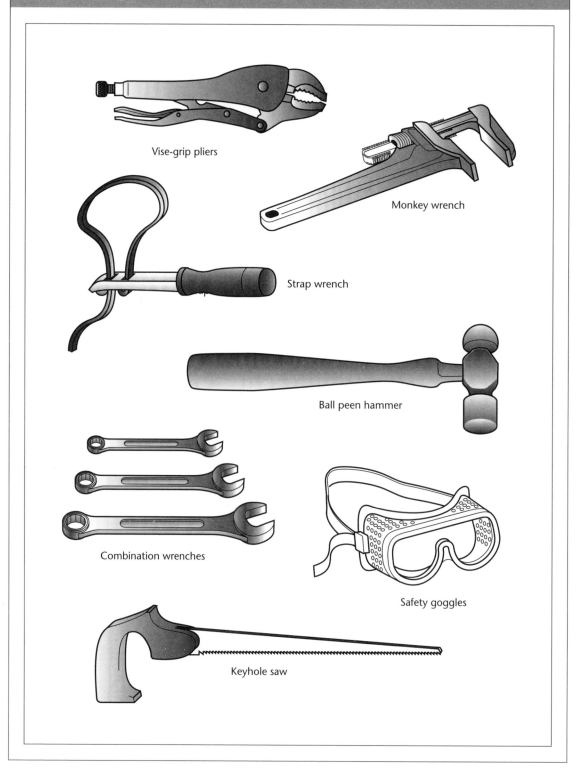

Vise-grip pliers

Monkey wrench

Strap wrench

Ball peen hammer

Combination wrenches

Safety goggles

Keyhole saw

every tool or material you might require for a plumbing project, but this basic tool-box will be necessary for most plumbing projects. If you take on a large project, make a project sketch and take it along to your hardware dealer. If you are replacing parts, remove the old part and take it along to the hardware store to make sure you get an exact duplicate. For seldom-used tools, your hardware dealer may loan them; if not, go to a tool rental store.

CHAPTER 2

Plumbing Safety

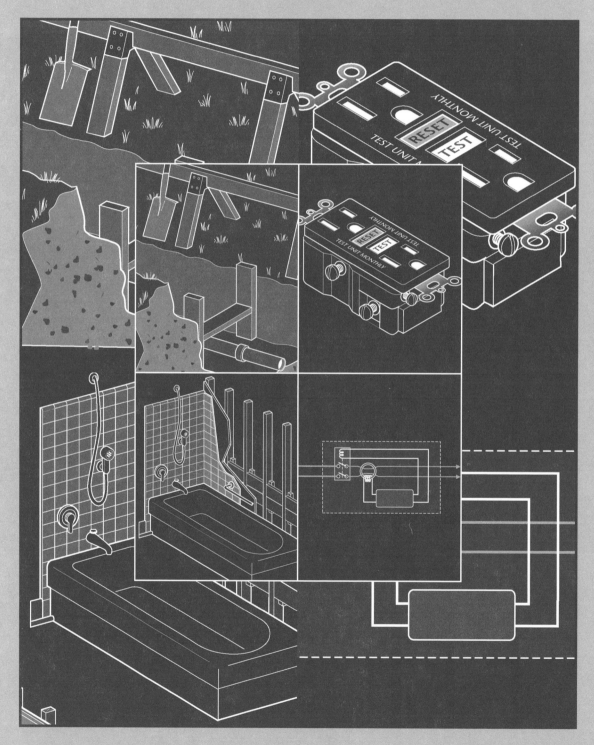

Throughout this book we have emphasized using safe procedures to avoid injury to the handyperson and fire from careless use of a plumbing torch. In this chapter we will review these safety tips, and examine another plumbing hazard: the open pipe trench.

Most of us will never undertake such an extensive project as connecting our own sewer and water service, or plumbing our own septic system. Piping for these systems involves working in a trench, often with our entire bodies below ground level. Trenches for city water and sewer hookup may be as deep as 15 feet (4.56 meters). Dangers from these trenches include cave-in of the earthen trench walls, which can injure or even bury the worker. Another possible hazard is playful children, curious spectators or workers falling into the trench. To avoid these possible injuries, observe the following tips.

First, never attempt projects that are far beyond your experience level. Earthen trench walls can suddenly collapse and bury you in the trench, sometimes with fatal results. Trench walls should be braced with shoring. You can rent hydraulic shoring, or make your own from 2 × 6s. To make wood shoring, set 2 × 6s against the walls on either side of the trench, then cut lengths of 2 × 6 long enough to provide cross-bracing between the two 2 × 6s. This shoring should be repeated every four feet (1.23 m) along the entire length of the trench. (See "Trench Safety" Illustration on page 18.)

The next step for trench safety is to provide barriers along the edges of the trench. These barriers may be sawhorses with planks between them, posts and lumber installed to form a fence, or visible wide tape such as that used by police at crime scenes. These barriers alert passersby that a hazard exists and prevents them from accidentally stepping into the trench in darkness.

Lastly, any construction site is a magnet to neighborhood children. Children like the sound of the backhoe or tractor, and trenches offer new play opportunities and worlds to explore. If you have open trench construction at your house, warn nearby parents and their children that the trench is hazardous and not a playground. No construction season goes by without news reports of children injured or buried in trench mishaps. Keep an eye on any trench construction in your area.

Plumbing and Electricity

Metal plumbing pipes offer a perfect ground to errant electric current. Never use corded electrical tools when working on plumbing. Instead, use battery-powered hand tools to avoid electrical shock.

Electricity and water don't mix. That is why modern codes require that all electrical receptacles installed near sources of water have ground fault circuit interrupters (GFCIs). (See "Ground Fault Circuit Interrupter" Illustration on page 19.) GFCIs are required for receptacles in kitchens, bathrooms and laundry areas, as well as outdoor receptacles.

A ground fault circuit interrupter is a device that measures the current flow in the hot (black) wire and the white (neutral) wire. The current flow for both these wires should be the same. If the current in the white wire falls below that of the black wire, it indicates that the current is going to ground. The GFCI will sense a current drop of as little as 40 milliamps. This current drop trips the GFCI's interrupter to instantly shut off the current before a dangerous shock can occur. (See "How the Interrupter Works" Illustration on page 19.) Electrical fuses or circuit breakers alone do not offer this level of protection against electrical shock.

If you do not have GFCI protection at receptacles

Trench Safety

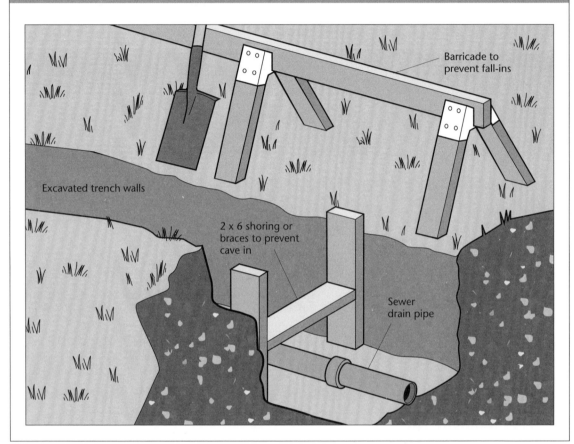

Barricade to prevent fall-ins

Excavated trench walls

2 x 6 shoring or braces to prevent cave in

Sewer drain pipe

near plumbing fixtures, you can install them yourself. The installation is similar to installing any new receptacle; directions are provided with each GFCI. Note that some GFCIs protect only the single receptacle where they are installed, while others also protect all other receptacles downstream on the same circuit.

Plastic Plumbing

Plastic plumbing is not only inexpensive, it is flexible and can be routed through framing or other obstacles without the need to make many joints. This not only reduces work time, but eliminates the cost of many fittings that would be necessary with metal pipe. From a safety standpoint, solvent-welded plastic pipe is safer as it eliminates the need for a soldering torch and use of an open flame to connect the pipe. No matter what your plumbing project there are plastic fittings available for any repair.

Bathroom Safety

In Chapter 9, Remodeling the Bath, we discuss how to build a safer bath, including installation of supports or grab bars to prevent slips and falls in the bath, which is the primary cause of bathroom injuries. Other bathroom safety tips: to prevent accidents and mistakes in taking prescription medicine, increase bathroom lighting for the elderly. To prevent electrical shock, install GFCI receptacles in the bathroom.

Ground Fault Circuit Interrupter

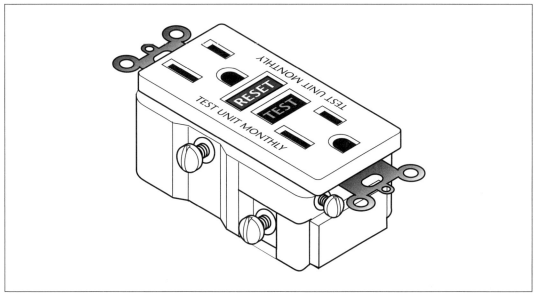

A GFCI receptacle has a three-hole or grounding plug. Fuses or circuit breakers will not react quickly enough to prevent a shock; thus the need for the quick-acting GFCI.

How the Interrupter Works

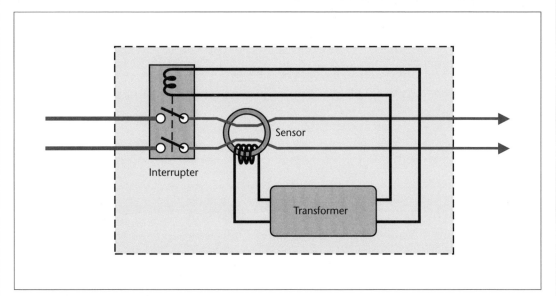

As current flows through the interrupter and transformer, current measured in both hot and neutral wires must be equal. If current in the hot wire is greater the sensor trips the interrupter to shut off the current.

CHAPTER 3

Understanding the Plumbing System

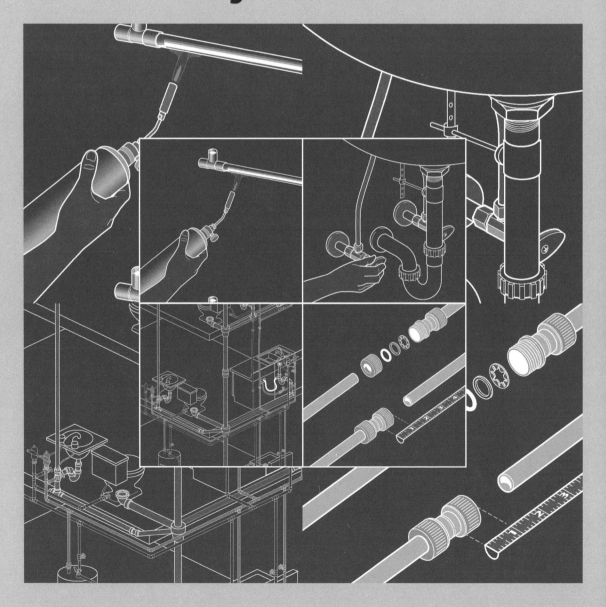

Your home plumbing consists of two separate systems, the water service system and the waste disposal DWV (drain-waste-vent) system. The water service system transports potable water from a water treatment plant through a water main and into the home. Potable water is water that is fit for human consumption and use. The piping is split into two layouts. One carries water directly to the water softener, if you have one, then to the water heater, and through hot water pipes to the various fixtures where hot water is needed. Like the hot water system, the cold water branch routes the water to various fixtures, including the sinks, toilets, bathtubs and laundry equipment, and also to exterior faucets called hose bibs or sillcocks.

The second part of the plumbing system is a drain-waste-vent system that carries waste and polluted water out to a sewer system, then to a waste plant, where the water is treated to remove solids and kill harmful bacteria. The treated water exits the treatment plant and is disposed of, usually in a body of water such as a river or an ocean.

If you want to avoid high plumbing repair bills and do your own maintenance, it is important to understand how both the water service and drain-waste-vent systems work. (See "Home Plumbing" Illustration on page 22.)

WATER SUPPLY SYSTEM

At the water service entry, you will find a water meter and a main shutoff valve. The water meter, of course, measures water usage for billing by the utility company. The gate valve allows you to turn off the water flow to the entire house, in case of emergency or to make needed repairs. There is another shutoff valve beneath the toilet tank or water closet. This valve permits you to shut off the water to service the toilet without having to shut off the water to the entire house. In older houses this may be the only fixture shutoff valve and you may have to shut off the entire system at the main shutoff valve to work on faucets in other fixtures. In modern houses, there are often service shutoff valves at sinks and laundry equipment, so you can shut down the water supply fixture by fixture instead of having to shut down the entire house. (See "Fixture Shutoff Valve" Illustration on page 23.)

Because the water supply pipes are pressurized, they can be smaller than drain-pipes, which operate by gravity. The pressurized supply pipes and fixtures are naturally more subject to leaks, due to the pressure. If pressure on the street supply pipe is over 80 pounds per square inch (psi) there is a pressure-reducing valve at the water entry. The water pressure in your house is 80 psi or less.

The water service pipe from the meter is usually ¾-inch (19.05 mm) in diameter. This ¾-inch pipe runs to the water softener, if any, and to the water heater. Branch pipes, ½-inch (12.7 mm) in diameter, are then run to each fixture where cold and/or hot water is needed. Depending on the amount of water needed at a given fixture, ½-inch (12.7 mm) or ⅜-inch (9.5 mm) pipes carry the water in the risers, the vertical pipes that extend up to the faucets and fixtures. Toilets and sinks permit ⅜-inch (9.5 mm) diameter risers; all others are ½-inch (12.7 mm)

THE DWV SYSTEM

The drain-waste-vent system includes drainpipes, sized to carry the wastewater load generated by a given fixture or appliance. In older houses, the branch drains to fixtures tend to be 1 ¾-inch (4.44 cm) steel pipes. The soil stack is the drainpipe for the toilet and, because it car-

Home Plumbing

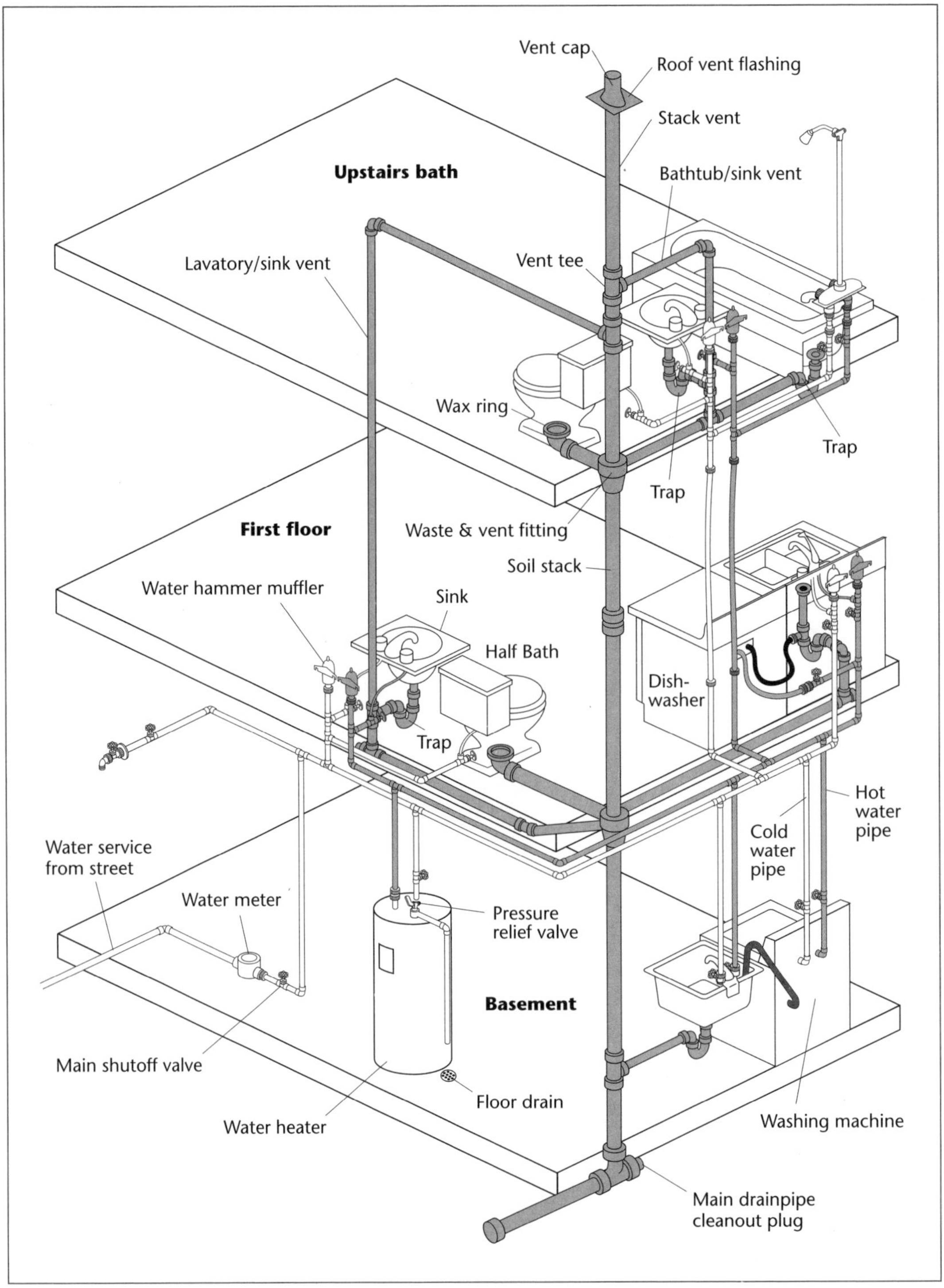

Vent cap

Roof vent flashing

Stack vent

Upstairs bath

Bathtub/sink vent

Lavatory/sink vent

Vent tee

Wax ring

Trap

Trap

First floor

Waste & vent fitting

Soil stack

Water hammer muffler

Sink

Half Bath

Dish-washer

Trap

Hot water pipe

Cold water pipe

Water service from street

Water meter

Pressure relief valve

Basement

Main shutoff valve

Floor drain

Washing machine

Water heater

Main drainpipe cleanout plug

ries away human waste as well as wastewater, is the largest drain line in the house. The smaller drainpipes feed into the soil stack. The cast-iron soil stack is 3 ½ inches (8.92 cm) in diameter. Plastic soil stacks may be 3 inches (7.6 cm) in diameter, but this is generally the minimum size allowed by code in most U.S. states. In cold weather, moisture may form ice in the stack vent above the soil stack, so the pipe size must be large enough to prevent complete freeze up and closure of the vent.

The cast-iron main drainpipe exits the basement and runs underground to join the main sewer at the street. Because the drain system is not pressurized and operates only by gravity, a slope of ¼-inch (6.35 mm) per running foot (30.48 cm) is maintained throughout the drainpipe system.

TRAPS

The toilet stool incorporates an integral trap; thus there is always standing water in the toilet bowl. In the drainpipe beneath all other fixtures there is a removable pipe that is shaped like a *U*, a *P*, a *J* or an *S*. These pipes are called traps, because when the fixture is used residual water is always trapped in this pipe.

The primary purpose of

Fixture Shutoff Valve

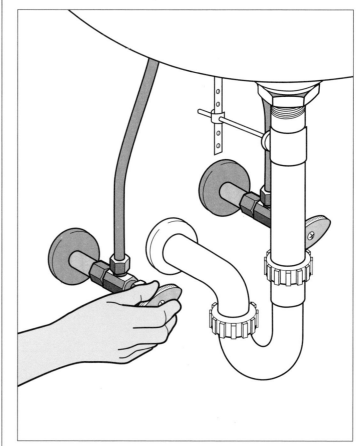

Under sinks and other fixtures you will find a service shutoff valve. This permits you to shut off the water to one fixture without shutting off the main shutoff valve. Note the trap at right: slip nuts can be loosened for trap replacement or auger access to plugged branch drain.

the trap is to hold water, to block sewer gases and odors from flowing back into the living space. The traps may be made of chromed steel or plastic. Steel traps are subject to eventual rust-out and leakage, because they are continuously filled with water. Replacing these traps is easy because they have large slip nuts on either end. If you must replace one, take

the old one along to the store to make sure you buy a matching replacement. The slip nuts are easily loosened by hand or by using channel-type pliers.

In addition to blocking sewer gases, drain traps serve a secondary function: the trap is the last defense against stray items entering the drainpipe and causing a blocked drain. If an item drops into the drain, remove the trap to remove any blockage. This can be rewarding. One landfill operator received a load of discarded building materials from a remodeling job on an old mansion. When he moved the bathroom sink, a large diamond ring dropped out of the trap. If you lose a ring in the sink, check the traps underneath.

In addition to checking the traps if a minor drain plug develops, there are cleanout plugs in horizontal drainpipes. By removing the cleanout plug, you can insert a rotary auger into the drainpipe and remove the plug material to free the drain.

VENTS

Nature abhors a vacuum. In order for water and waste to leave the drain system, air must enter behind the water. This principle can be observed by placing a drinking straw into a glass of liquid. If you seal the top end of the straw with your finger, the liquid will remain in the straw when you pull the straw from the glass. You have, in fact, created a vacuum by blocking the air entry. This is why vent pipes are needed in your DWV system. Unless air pressure equal to ambient pressure is maintained in the pipes, the water cannot run out. The waste pipe can double as a vent, assuming it is short or large enough to carry away the wastewater without completely filling the pipe. The minimum waste pipe diameter should be at least 1 ½ inches (2.86 cm).

The main vent is the stack vent, located directly above the soil stack of the toilet. The stack vent extends through the roof and may terminate in a vent cap that prevents rain, snow or debris from entering the stack. The stack vent is the same diameter as the soil stack. All branch waste pipes empty into, and are vented by, the soil/stack vent.

If a fixture's drainpipe enters the stack below the toilet drain, the water flushed from the toilet may siphon the water from that fixture's trap. In this case, a branch vent pipe, or revent, must be installed. If a fixture is too far away from the vent stack, a separate vent pipe must be installed through the roof.

CHAPTER 4

Working with Water Pipe

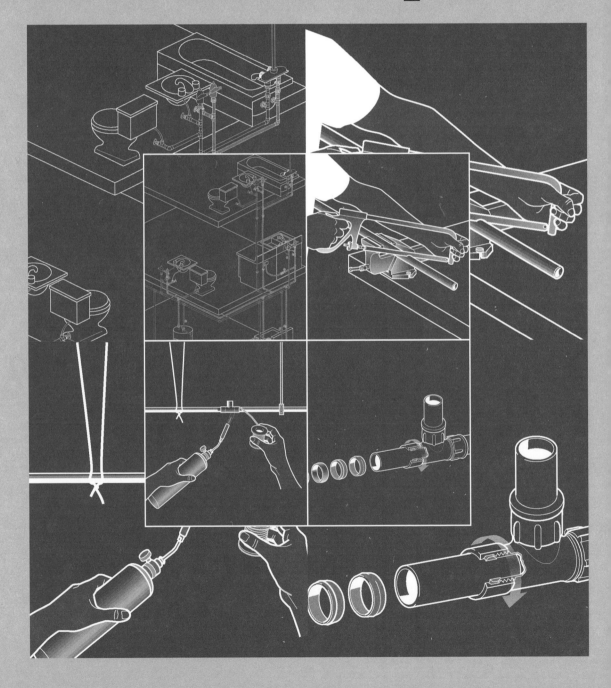

Until the early 1960s water supply pipes were made from galvanized steel, and drainpipes from cast iron. The plumber set up a pipe vise in the living room of the new house. As the longest room in the house, the living room was long enough to accommodate the pipes that he would cut and thread. The plumber then spent several days cutting and threading pipe, thus making the house interior inaccessible to other building trades until the job was done.

This process was not only a lengthy one, but during the threading process half of the steel pipe thickness was cut away. The threaded area where fittings such as Ts and Ys were installed was subject to rust-through, so leaks often developed later at the pipe joints.

Because the cutting and threading process required expensive tools, steel pipe installation was not a handyperson-friendly job, so plumbing was left to professionals. During the 1960s, use of steel water piping was abandoned in favor of copper pipes. Two plumbers might show up in the morning, cut and solder the workable copper, and finish the rough-in or pipe fitting before the day was over. Copper pipes not only sped up the work, but they were resistant to corrosion and future problems.

A few years later plastic pipe and fittings became available. Both copper and plastic made possible do-it-yourself plumbing. (See "Water Service System" Illustration on page 28.)

STEEL PIPE

As noted above, steel pipe in houses gave way to copper about forty years ago. If your water pipes are steel, you may find an independent hardware store that will cut and thread the pipe to make repairs. A simpler solution is to replace the damaged steel pipe section with plastic repair pipe and fittings. Ask your hardware dealer to help you select the appropriate steel-to-plastic fittings.

COPPER PIPE

Copper pipe is available either as soft rolled tubing or in 10-foot-long (3.048 m) rigid pipes, and is available in a variety of diameters. Three-quarter-inch (19.05 mm) pipes are used to connect the water main to the house, and continue inside the house to the water softener, if any, and to the water heater. Check your local code: some areas may require that 1-inch-diameter (25.4 mm) pipe be used at the water service entry. Branch pipes to plumbing fixtures are usually ½ inch (12.7 mm) in diameter; risers may be ⅜ inch (9.5 mm) in diameter.

The rolled copper tubing is used to connect the main water supply to the meter at the house. The long rolled tubing is required because it can be laid in the excavated trench without making joints that will be buried, which could result in an inaccessible leak, when the earth is pushed back into the trench. You may also choose to use the flexible tubing in appropriate diameters when plumbing a remodel job. The tubing can be bent and routed through obstacles where it would be difficult to install rigid copper pipes.

If you are doing an outdoor piping job, a few precautions are in order. First, check with your local building department to learn the depth requirements for burying water pipes in your area. The frost depth—the maximum depth to which the ground is frozen in winter—varies greatly from one region to another. Water pipes that are not laid at the proper depth may freeze and burst in cold weather.

Another tip for burying pipe: when the connections are secure, shovel clean soil in by hand until the pipe is covered. Then use a tractor to fill the trench. If a tractor is used to push soil onto bare pipe, large rocks or chunks of soil can dent or penetrate the pipe, causing a leak in

Water Service System

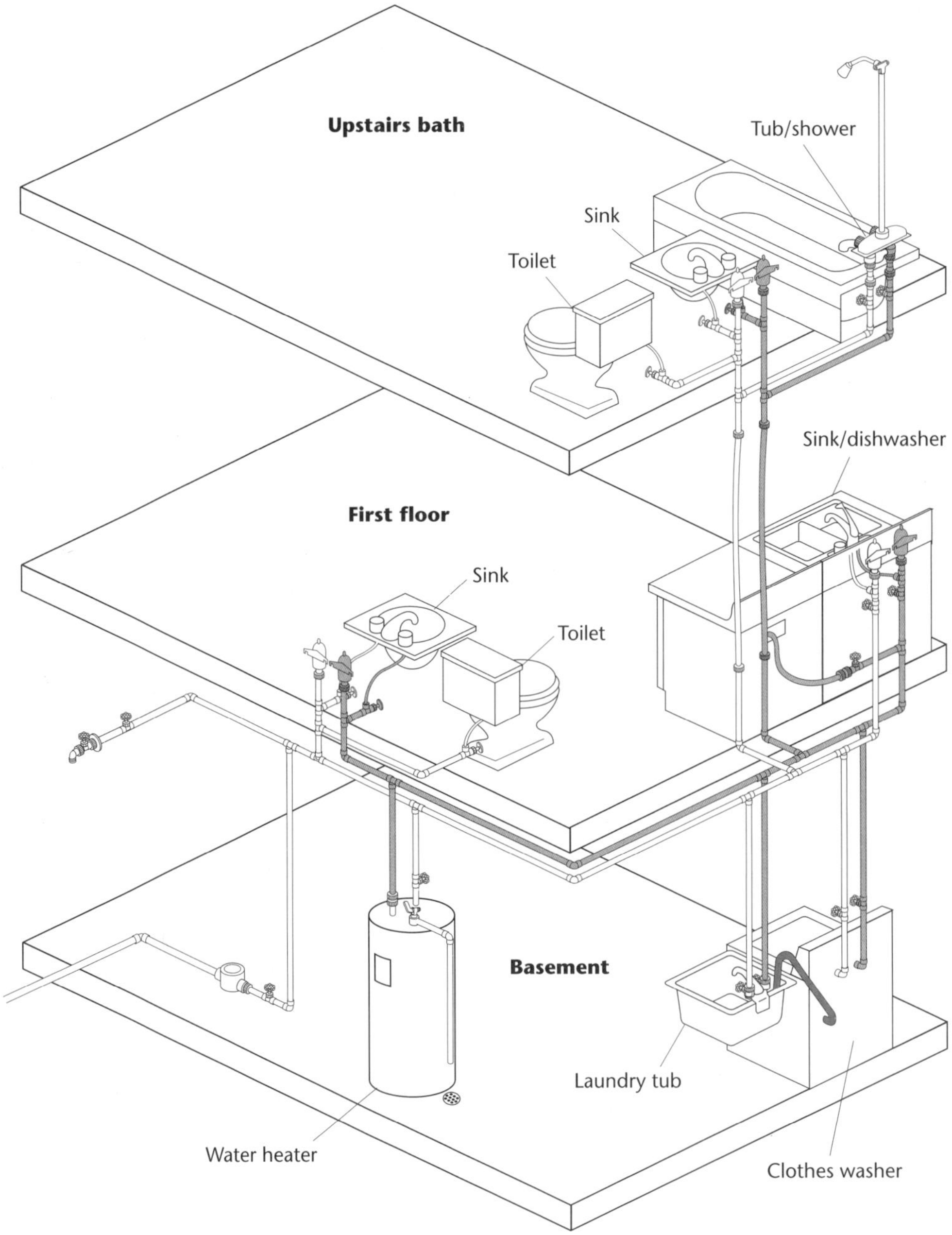

Upstairs bath

Tub/shower

Sink

Toilet

Sink/dishwasher

First floor

Sink

Toilet

Basement

Laundry tub

Water heater

Clothes washer

the now-buried pipe. These leaks are expensive to repair.

CUTTING COPPER PIPE

When measuring water pipes for installation, don't neglect to add the length that the pipe will extend into the fitting to the total length of the pipe. This is called the makeup. To determine how much to add, measure the depth of the socket on the fitting to the hub or shoulder; then double the number because there will be a fitting joint at each end of the pipe. Add this figure to the length of the pipe between fittings to get the length you must cut.

To ensure leak-free joints, all pipe ends, whether copper or plastic, should be cut square, so that they sit squarely against the hub in the fittings. You can use a sharp hacksaw and an inexpensive plastic miter box to ensure a square cut in either plastic or copper pipe. If you need to make many pipe cuts, invest in a tool called a tubing cutter. The tubing cutter has a capacity to handle pipe from ¼ inch to 1 ½ inches (7 mm to 40 mm) in diameter. The copper tubing cutter resembles locking or vise-grip pliers, with an adjusting knob at the end of the handle and a cutting wheel on one of the jaws.

To make a cut, position the pipe in the jaws of the cutter and adjust the knob until the pipe is firmly positioned against the cutting wheel. Turn the tubing cutter several rotations around the pipe. Re-tighten the cutting wheel and turn again. Repeat until the pipe is cut completely through. (See "Cutting Copper Pipe" Illustration 2 on page 30.)

When the pipe is cut, there will be burrs from a hacksaw or a slight ridge from the tubing cutter on the inside of the pipe opening. If the burrs or ridge are not removed, the pipe may vibrate or make a whistling noise when water runs through it. There are two deburring blades on the backside of the tubing cutter. Fold the tool blade out and ream out the burrs or ridge at the pipe opening. If you do not use the tubing cutter, you can use a small metal file to deburr the cut end of the pipe. (See "Cutting Copper Pipe" Illustrations 3 and 4 on page 30.)

CLEANING COPPER PIPE AND FITTINGS

To remove any oxidation from the pipe and fitting socket, the inside of the fitting socket and the pipe end must be cleaned. For limited work you can use fine emery cloth or steel wool to clean the pipe and fittings. (See "Cleaning Copper Pipe" Illustrations 1 and 2 on page 31.) Make sure you clean the inside of the fitting socket completely. For about $10, you can buy a wire fitting brush that cleans ½-inch or ¾-inch (12.7 mm to 19.05 mm) pipe. The wire fitting brush has two male brushes for cleaning inside the pipe fittings and two female brushes for cleaning the exterior of the pipe. (See "Cleaning Copper Pipe" Illustration 3 on page 31.) Be sure to clean the pipe beyond the point where it joins the fitting, to ensure a good bond between the copper fitting and the solder. This cleaning step is vitally important. Solder will not bond to oxidized surfaces.

BENDING COPPER PIPE

If you are using copper tubing for a remodeling project, there will be a point at which you must make a bend in the tubing. The challenge is to make the bend without kinking or collapsing the tubing. There are several ways to bend the tubing without damaging it. You can use an electrician's tool called a hickey to bend the pipe. You can rent this at most tool rental outlets.

Another bending option is a spring tubing bender. These come in various diameters and are also available at tool rental stores. To make a

Cutting Copper Pipe

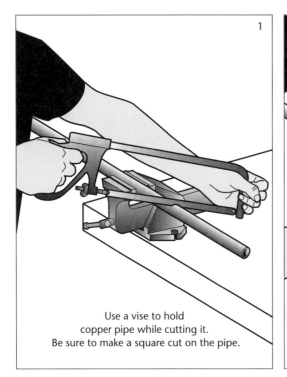

Use a vise to hold
copper pipe while cutting it.
Be sure to make a square cut on the pipe.

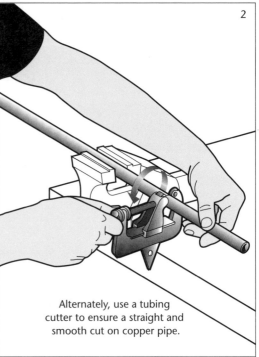

Alternately, use a tubing
cutter to ensure a straight and
smooth cut on copper pipe.

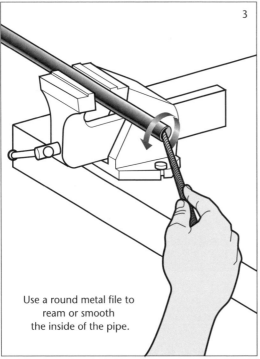

Use a round metal file to
ream or smooth
the inside of the pipe.

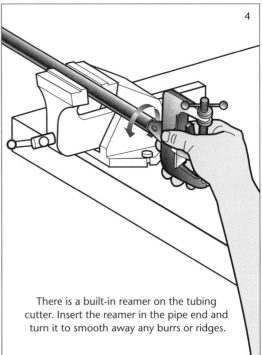

There is a built-in reamer on the tubing
cutter. Insert the reamer in the pipe end and
turn it to smooth away any burrs or ridges.

Cleaning Copper Pipe

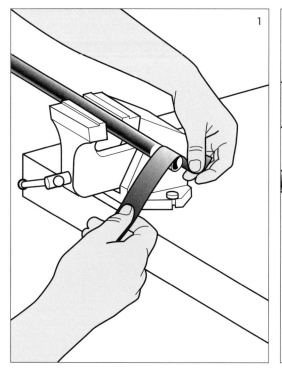

1

2

Use fine emery paper or steel wool to clean the pipe end and the inside of the fitting socket.

This will remove any oxidation and ensure a strong solder bond between the fitting and pipe.

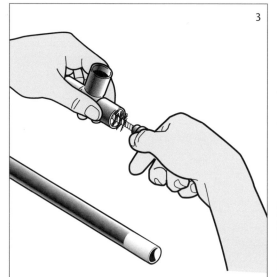

3

A wire fitting brush makes cleaning pipe and fittings an easy task, and ensures a bright copper surface, which will accept the joint solder.

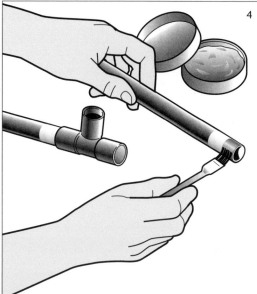

4

Use a brush to apply flux to both the pipe end and the inside of the fitting socket. Do not touch the cleaned surfaces with your fingers.

Applying Heat

With the solder torch, apply
heat first to the pipe...

...then to the fitting.
Apply heat evenly around
the pipe and fitting.

bend, you simply slip the spring coil over the tubing to the point at which you wish to make a bend. You can then bend the tubing over your knee while the spring coils prevent the tubing from distorting and kinking. Once you've made the bend, just use a twisting motion to pull the spring off the tubing. (Refer to "Bending Copper Tubing" Illustration on page 35.)

To make sharp bends in copper tubing, fill the tube with dry sand and plug the pipe ends. Make the bend. The sand will permit you to do this and will support the walls of the tubing so they cannot collapse inward. Once you've made the bend, use a garden hose to flush any remaining sand out of the tubing.

SOLDERING COPPER PIPE

Soldering copper pipe is a skill that amateurs may find difficult to acquire at first, but a little practice soon has do-it-yourselfers sweating perfect solder joints. There are a few suggestions that should be followed to ensure perfect, watertight solder joints. The first step, cleaning the pipe and fittings, we have discussed above. To ensure a watertight solder joint, the copper surfaces to be joined must be bright as a new penny.

Soldering Copper Pipe

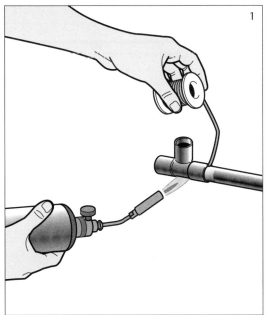

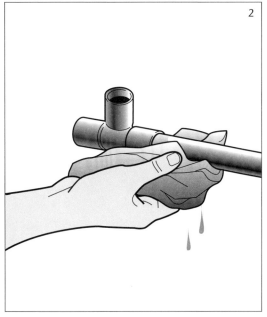

Unroll a length of solder wire and test the joint by touching the solder to the surface. If the solder melts, move the solder tip around the joint to ensure an even application of solder and a watertight joint.

Use a cold wet cloth to smooth and clean the solder joint.

The next step involves using a brush to apply solder flux to the surfaces to be soldered. Soldering flux is a combination of petroleum jelly and mild acid, which is used to prevent oxidation from reforming on the copper before or during the soldering process. You can buy solder with an acid core; this product is useful for soldering copper wires or small copper projects. For soldering pipes, use the flux that is packaged in a tin container in paste form.

Dab the brush in the paste flux and apply the flux to the surfaces to be joined.

Coat the cleaned pipe end, then insert the brush into the fitting hub and coat the interior surfaces of the fitting with flux. When handling the pipe and fittings, be careful not to touch the soldering surfaces with bare fingers. The oil from fingerprints can contaminate the solder surfaces in such a way that the solder will not flow and fill the joint between the pipe and fitting. (See "Cleaning Copper Pipe" Illustration 4 on page 31.)

Solder is an alloy, or a combination of metals, that have a low melting point,

usually a combination of lead and tin. In the past the common lead-tin solder ratio was 50:50, but solder with a high lead content was banned in 1984 as it was deemed a health hazard. Officials feared that the lead in the solder would leach into drinking water and cause health problems. Now solders may be 95 percent tin to 5 percent lead, safe for home water systems.

It is easier to solder when the work can be done at bench-top level, rather than soldering overhead. Assemble and solder as many joints as possible at bench

Pipe Hangers

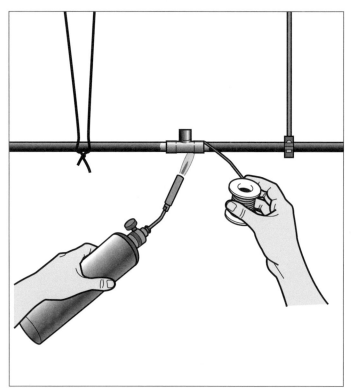

Use temporary (left) or permanent (right) wire hangers to support pipe assemblies while you join them together overhead.

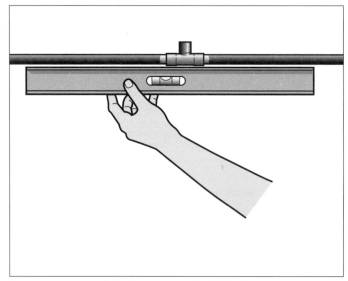

Then use a carpenter's level or square to check that the pipe joints are straight.

top level, then lift the assembly into place. Use pipe hangers to hold the pipe assembly in position until the pipe joints are all soldered. It's important to use pipe hangers of the same metal as the pipe you are working with; that is, use galvanized steel pipe hangers for steel, copper pipe hangers for copper. This is crucial because a reaction can occur between any two unlike metals, causing a corrosion process. Plastic pipe hangers, of course, can be used with either plastic or metal pipe, because plastic contact with metal causes no reaction. When hanging pipes, make sure that horizontal pipes are hung level, and vertical pipes are plumb.

Before lighting the torch, wet an old towel with cold water and wring it out. Place the wet cloth near the solder project, within easy reach. Light the soldering torch and apply the flame to the pipe, near the fitting. This will bring the pipe to soldering temperature more quickly than will applying the flame to the fitting alone. (See "Applying Heat" Illustration on page 32.) Heat the pipe next to the fitting until the flux melts and becomes a clear liquid. After preheating the pipe end, move the torch so the flame is applied to the fitting. Move the torch around the fitting so all sides are evenly heated. After a

Bending Pipe

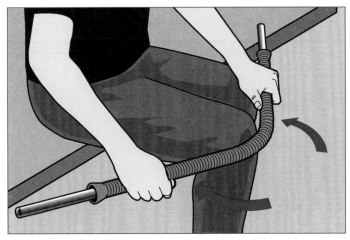

Slip a spring tubing bender over the pipe so it is centered over the point at which you want to make the bend. Bend the tubing over your knee. Remove the spring bender by using a pulling and twisting motion.

Tip Box

Soldering a pipe that has water in it can be a frustrating task. When it is heated any water droplets in the pipe will become steam, blowing back the molten solder in the joint. To solder a pipe that has water residue in it, roll a piece of white bread into a ball and insert the bread into the end of the pipe. The bread will absorb any moisture in the pipe and you will then be able to solder the joint successfully. When you turn the water on, the bread will soften and flow out through the faucet.

If you need to take apart a soldered joint, turn off the water and open nearby faucets to drain them. Always work with a dry pipe. If you attempt to soften the old solder joint and the solder will not melt, there may still be water in the pipe. If there is, the water will absorb the heat from the solder torch and the solder will not become hot enough to melt. If this is the case, open all faucets near the pipe you are working on. If this fails, you may have to use a tubing cutter to cut the pipe and drain off the water, then solder in a new pipe section. Note: some connections such as air conditioning connections may be silver soldered. In this case a propane torch will not be hot enough to melt this type of solder. Have a pro solder these joints.

couple of seconds, apply the solder to the joint. If the solder does not melt at all, or melts slowly, pull it away and continue to apply heat to the fitting. Touch the solder to the joint again. When the solder melts, quickly apply it in a circular movement around the entire joint. (See "Soldering Copper Pipe" Illustration 1 on page 33.) If the joint is the proper temperature, the solder will flow into and fill the gap between the pipe and the fitting. You will see the flux flow out of the joint as a clear liquid as the solder flows into the joint.

Now, turn off the torch and set it down, grasp the wet cloth and wipe around the solder joint while the solder is still molten and flowing. Caution: the pipe and fitting are very hot; to avoid burns, wrap the wet cloth around your hand. This procedure will wipe away the flux and smooth the solder joint, eliminating any pinhole leaks and creating a professional-looking solder joint. (See "Soldering Copper Pipe" Illustration 2 on page 33.) Tip: if you are soldering a fitting such as a T where the pipe joints are close together, use the wet cloth to wrap the joints that are already soldered. This will keep the soldered joint cool and intact while you are soldering any nearby joint. (See

Multiple Solder Joints

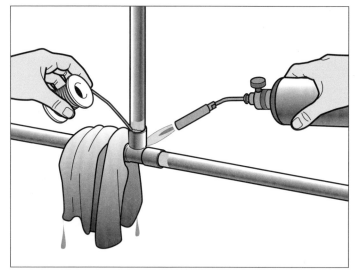

When soldering two or more joints that are close together, wrap the joint you have just soldered with a cold wet cloth to cool it while you solder the next one.

"Multiple Solder Joints" Illustration on page 36.)

PLASTIC PIPE AND FITTINGS

Plastic pipe and fittings are easy to work with, a real advantage for do-it-your-selfers. Plastic pipe assembly is easy. Because no open flame is needed to assemble plastic, unlike copper, there is no danger of fire or burn injury to the worker. Plastic pipe and fittings can be used for repairs that involve replacing copper pipes and offer an easy solution for the new homeowner's new plumbing projects.

Be sure the plastic pipe and fittings you buy can be used for either hot or cold water lines. Choose poly-butylene (PB) flexible piping for projects where rigid piping may be difficult to install. Polybutylene plastic cannot be solvent welded and must be joined by mechanical fittings. These fittings are available at your dealer. (See "Polybutylene Pipes and Fittings" Illustration on page 40.)

Select chlorinated polyvinyl chloride (CPVC) products if you need rigid pipe. CPVC products can be used either for hot or cold water piping, and can be solvent-welded. As with all pipe, CPVC pipe expands and contracts in response to changes in water tempera-ture. Allow for a ½ inch (12.7 mm) of expansion on

a 10-foot-long (3.05 m) pipe. Also, use plastic pipe hangers, spaced 32 inches (0.81 m) apart to support the piping while permitting free expansion and contraction of the pipe.

Plastic pipe is easy to cut, using a plastic tubing cutter or a fine-tooth saw such as a hacksaw. Use a sharp knife to clean any burrs from the cut end of the pipe. (See "Joining Plastic Pipe" Illustrations 1 and 2 on page 37) To join CPVC pipe to fit-tings, use a solvent-welding process. First use the pipe manufacturer's solvent cleaner to clean both the end of the pipe and the socket of the fitting. Then use solvent cement to per-manently join the pipe and fitting. Apply the cement with a brush, putting a gen-erous coat on both the pipe end and the inside of the fit-ting socket. (See "Joining Plastic Pipe" Illustrations 3 and 4 on page 37.) Insert the pipe into the socket with a slight twisting motion to seat the pipe in the proper position. Let the unit sit for a few minutes before moving it. (See "Joining Plastic Pipe" Illustration 5 on page 37.)

PLUMBING FITTINGS

Perhaps the most difficult part of doing your own plumbing is being able to lay out a job and select the

Joining Plastic Pipe

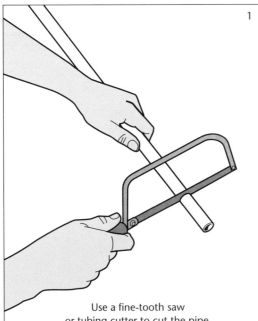

1

Use a fine-tooth saw
or tubing cutter to cut the pipe.
Be careful to make a square cut.

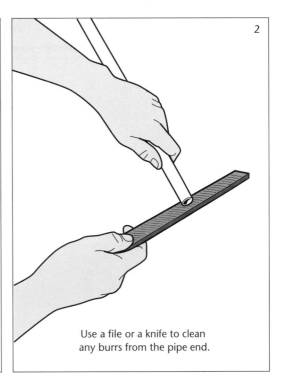

2

Use a file or a knife to clean
any burrs from the pipe end.

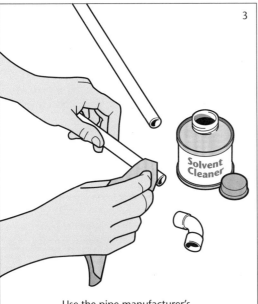

3

Use the pipe manufacturer's
solvent cleaner to clean the pipe
and the inside of the fitting.

4

After applying the pipe cleaner,
apply the solvent weld cement to
both the pipe end and the inside
of the fitting socket with the
brush supplied.

Joining Plastic Pipe

The solvent weld cement sets quickly, so insert the pipe end into the fitting socket with a twisting motion.

You must adjust the alignment of the pipe immediately, before the solvent cement sets.

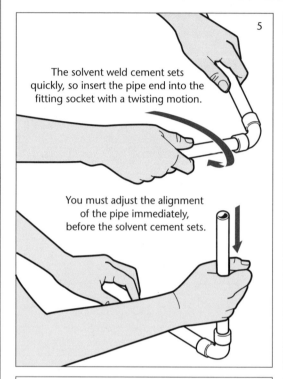

5

To connect CPVC pipe to copper, use a compression fitting and two-part adapter.

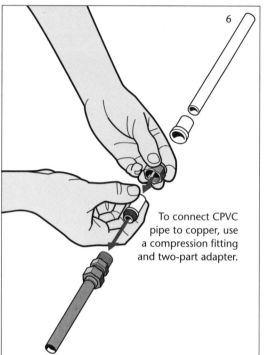

6

Use an adjustable wrench to tighten the compression fitting. Use a second wrench to hold the fitting body for final tightening.

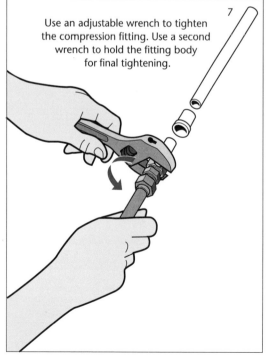

7

8

Solvent weld the two adapter parts together, then solvent weld the second part of the adapter to the CPVC pipe.

Polybutylene Pipes and Fittings

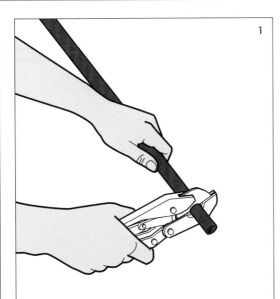

1

To ensure a straight cut,
use a plastic tubing cutter
to cut polybutylene (PB) pipe.

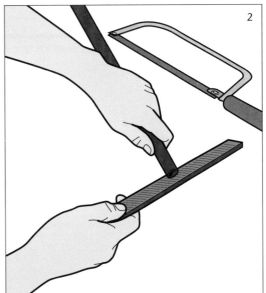

2

For limited cuts, you can use a
hacksaw or knife to cut the pipe. Clean
away any burrs using a file to smooth the cut.

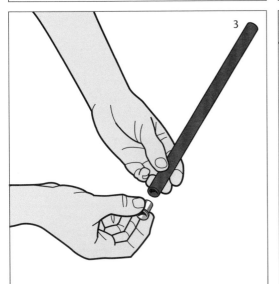

3

Some fittings come with a stainless steel
support sleeve. Insert the sleeve into the pipe
end to prevent it from being crushed
when it's inserted into the fitting.

4

You can use ordinary compression fittings
with PB pipe. Just slide the compression nut
over the pipe, then the compression ring.
Tighten the nut to seat and seal the ring.

Polybutylene Pipes and Fittings

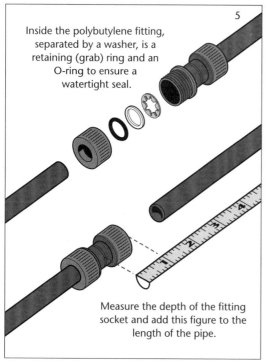

Inside the polybutylene fitting, separated by a washer, is a retaining (grab) ring and an O-ring to ensure a watertight seal.

Measure the depth of the fitting socket and add this figure to the length of the pipe.

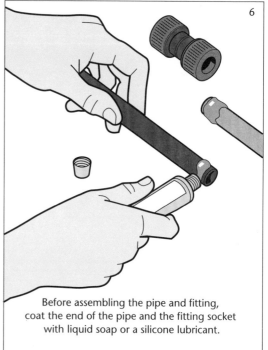

Before assembling the pipe and fitting, coat the end of the pipe and the fitting socket with liquid soap or a silicone lubricant.

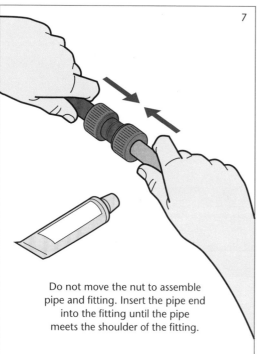

Do not move the nut to assemble pipe and fitting. Insert the pipe end into the fitting until the pipe meets the shoulder of the fitting.

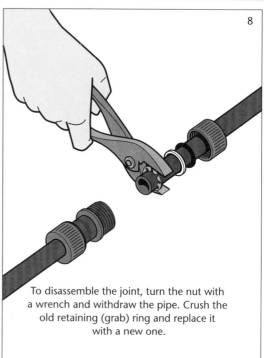

To disassemble the joint, turn the nut with a wrench and withdraw the pipe. Crush the old retaining (grab) ring and replace it with a new one.

Common Pipe Fittings

Flush bushing

Cap

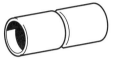

Coupling w/stop

Adapter

Cross fitting

45° L or elbow

Adapter

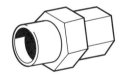

Union

45° street L or elbow

Adapter

Companion flange

90° L or elbow

Common Pipe Fittings

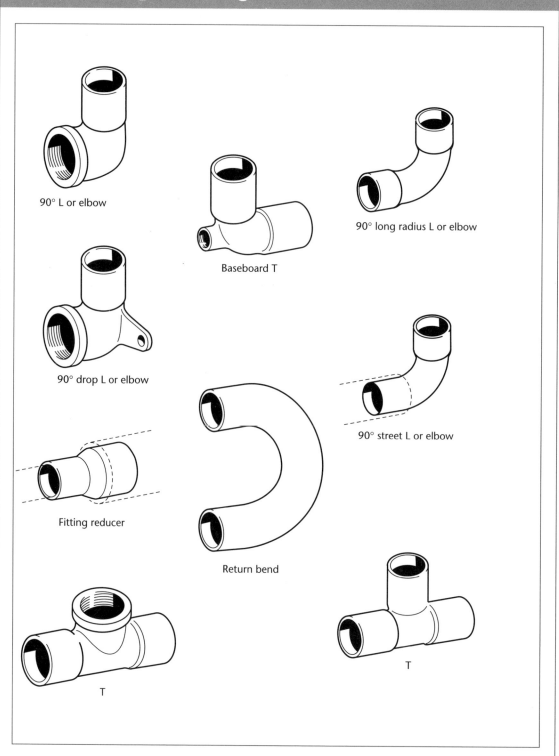

90° L or elbow

Baseboard T

90° long radius L or elbow

90° drop L or elbow

Fitting reducer

Return bend

90° street L or elbow

T

T

Valves

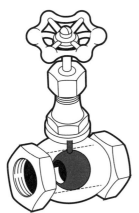

Globe valve

Gate valve

Stop-and-waste valve

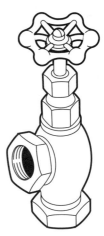

Angle valve

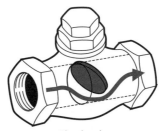

Check valve

How a Compression Fitting Works

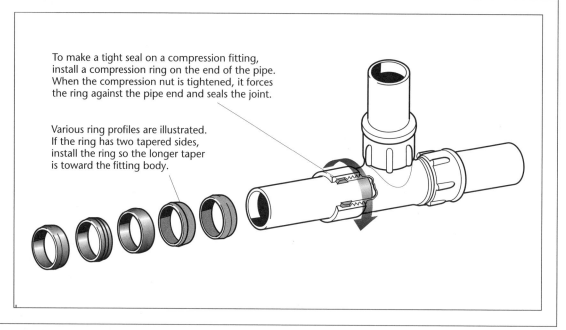

To make a tight seal on a compression fitting, install a compression ring on the end of the pipe. When the compression nut is tightened, it forces the ring against the pipe end and seals the joint.

Various ring profiles are illustrated. If the ring has two tapered sides, install the ring so the longer taper is toward the fitting body.

necessary fittings. Most home centers offer a bewildering array of parts bins, stocked with fittings whose labels indicate that you have chosen a "street L," an "adapter," a "union" or a "baseboard T." To complicate matters, customers often pick the wrong part out of a bin, examine the label, find it's the wrong size or wrong fitting and then throw it into the nearest bin. The result is that different kinds of fittings may be mixed up in the bin. You must always read the label of the individual package you choose to ensure you have the right fitting.

Running back to the store for missing parts can be a frustrating and time-consuming chore that can drag out a project until it seems endless. To reduce the frustration factor, make a working drawing of the project you are doing. Include the pipe size and the length of the pipe run. Draw in any branch lines, any change of direction and level (See page 23). Note what the pipe will be connected to, i.e., a fitting or valve. Take the working drawing along so that both you and the store clerk can consult it. If you are doing a repair job such as fixing a leaking faucet or replacing a valve, take the old part along to be sure you get a matching replacement part.

To help you learn the terminology of plumbing, we have included illustrations showing the most commonly needed pipe fittings. Note that Ts, Ys and Ls let you change pipe direction. Reducers let you change pipe size for a branch line; couplings let you join the ends of two pipes together. A union lets you join two pipes so they can later be separated for needed repairs or replacement of a fixture.

Note that some fittings are threaded and can be assembled with a wrench. The fitting with the threads on the exterior are called male fittings or MIP, which stands for male iron pipe. The socket they are screwed into, with threads on the interior, are called female fittings or FIP for female iron

Making a Compression Joint

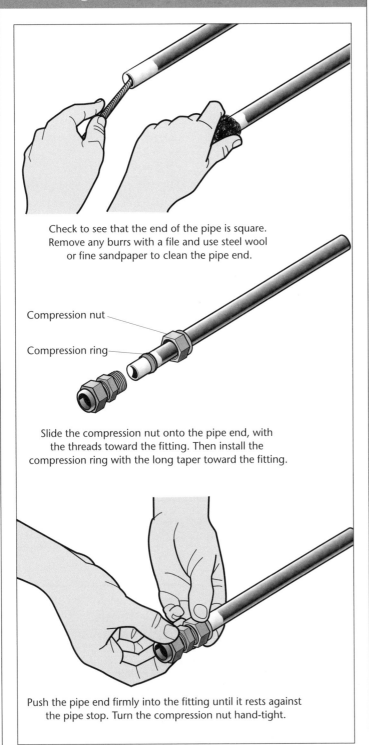

Check to see that the end of the pipe is square. Remove any burrs with a file and use steel wool or fine sandpaper to clean the pipe end.

Compression nut

Compression ring

Slide the compression nut onto the pipe end, with the threads toward the fitting. Then install the compression ring with the long taper toward the fitting.

Push the pipe end firmly into the fitting until it rests against the pipe stop. Turn the compression nut hand-tight.

pipe. Other fittings are not threaded and must be soldered (copper) or solvent welded (plastic pipe) together.

Pay special attention to the size of pipe you will be working with to be sure the selected fittings match the pipe size. As a rule, the water service entry pipe, hot and cold water mains, and water heater and water softener pipes are ¾ inch in diameter and therefore use ¾-inch fittings. Branch water service pipes to the bathroom, kitchen and laundry are ½-inch pipes, requiring ½-inch fittings. The risers or supply tubes to all fixtures, including the toilet, bathroom sink, laundry tub, kitchen sink and dishwasher are ⅜ inch in diameter.

Plastic drainpipes are 3 inches (7.6 cm); cast-iron drainpipes 3 ½ inches (8.9 cm) in diameter. If you are working with large-diameter drainpipes use no-hub connectors. No-hub connectors have a neoprene sleeve and two adjustable clamps. To connect two pipes, just slip the pipe ends into the neoprene sleeve and use a ratchet wrench to tighten the adjustable clamps.

A variety of valves is available to fit specified service needs. A gate valve allows full flow of water and is used for fixtures such as the water heater where there are few

Making a Compression Joint

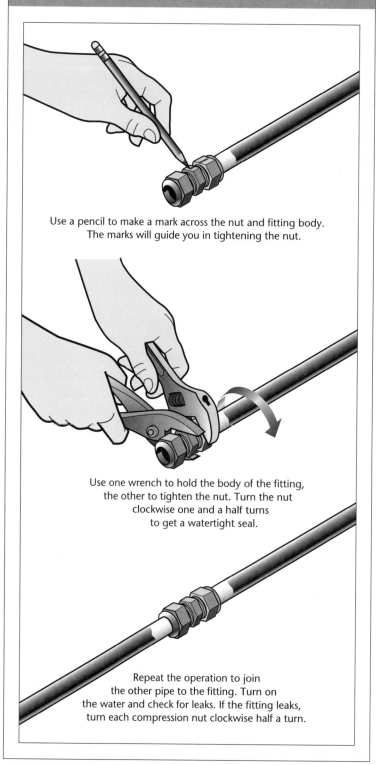

Use a pencil to make a mark across the nut and fitting body. The marks will guide you in tightening the nut.

Use one wrench to hold the body of the fitting, the other to tighten the nut. Turn the nut clockwise one and a half turns to get a watertight seal.

Repeat the operation to join the other pipe to the fitting. Turn on the water and check for leaks. If the fitting leaks, turn each compression nut clockwise half a turn.

occasions when the water must be turned on or off. Leave the gate valve fully open at all times. Globe valves are used for fixtures where you may wish to change the water flow rate. Globe valves can be left only partially open without damaging the valve. Use an angle valve when you want to change the direction of water flow. Stop-and-waste valves have a cap on one side that can be removed to drain the pipes beyond the valve. Check valves are used to prevent backflow, because water can flow in only one direction. To acquaint yourself with fitting and valve terms, refer to our fitting illustrations (See pages 41–43).

COMPRESSION JOINTS

Compression fittings allow you to make a pipe joint without soldering or solvent welding. With this kind of fitting, you can join two pipe ends using only a pair of wrenches, and to disconnect the joint at a later date all you need is a wrench. For example, flexible copper connector kits, used for connecting a new water heater, include compression fittings so you can disconnect the heater for service or replacement.

The compression fitting has metal rings, usually copper, which are compressed

How a Water Hammer Muffler Works

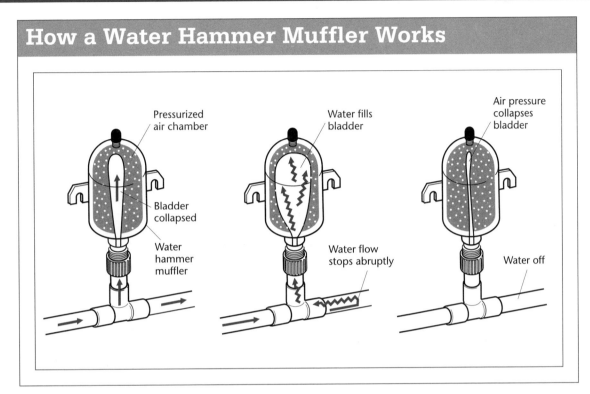

Pressurized air chamber

Bladder collapsed

Water hammer muffler

Water fills bladder

Water flow stops abruptly

Air pressure collapses bladder

Water off

against the pipe end as the nut is tightened to form a watertight seal. To assemble a compression joint, first slip the compression nut over the pipe end, with the threaded side of the nut facing the compression fitting. Then slip the ring over the pipe end. The ring may have two sloping or tapered sides. If one tapered side is longer than the other, install the ring so the long tapered side faces the fitting. (See "Making a Compression Joint" Illustration on page 45.)

NOISY WATER PIPES

If your water pipes are noisy, first check the pipe hangers. Metal or plastic water pipes expand and contract with changes in temperature. If the pipes cannot move freely in their hangers, they will make a noise as they rub against the floor joists. Check the hangers to see if they hold pipes too tightly against the joists. Remove the hanger and reinstall it so there is a slight space between the pipe and the joist.

WATER HAMMER

Unlike air, water cannot be compressed. Water used in plumbing moves along under pressure, so if water pipes vibrate or bang loudly when a valve shuts off abruptly you will hear water "hammer." This most commonly occurs when the fill valve snaps shut in a clothes washer or other appliance.

The way to silence water hammer is by installing a water hammer muffler. The muffler is a plastic bulb with a collapsible bladder inside and with air space between the outer shell or bulb and the bladder. When water flow stops abruptly, the water surges into the bladder, where it is cushioned by the air trapped between the bladder and the bulb. Installation of a water hammer muffler is easy, and instructions are included with the unit. The muffler can be used with special adapters so you can install it on ½-inch copper or plastic water pipe. (See Illustration above.)

CHAPTER 5

Drain Maintenance

The failure of the drain-waste-vent system can result in overflowing fixtures, water damage to home contents, possible damage to the home structure and a messy cleanup job. Many or most clogged drains are the result of failure to observe ordinary precautions to prevent drain problems. The best way to prevent drains from clogging is to take a fixture-by-fixture survey of the most common clog problems and how to avoid them.

TOILET

The most common type of drain clog is a plugged toilet. Keep in mind that the toilet stool is not a trash disposal. The toilet is meant to dispose of human waste only. Toilet tissue is designed to turn into to slush when wet, but other paper products such as soap wrappers can cause a clogged drain. Women's sanitary products will absorb water and expand when wet, creating a total drain blockage. Keep a wastebasket in the bathroom and use it to dispose of waste paper and sanitary products.

To prevent your child from throwing toys into the toilet, keep toys out of the bathroom and the lid down as a matter of routine.

BATHTUBS, SHOWERS AND SINKS

Bathtubs, showers and sinks are also prime locations for clogged drains. If you clean out these drains, you will often find that the culprits are soap scraps and human hair. Hair is especially common in clogged drains. It will not deteriorate, and has a spring-like quality that causes it to coil and lodge in the drainpipe. Install a sink or tub strainer at the drain outlets of these fixtures, so that hair washed away during shampooing will not lodge in clumps and block the drains.

Soap scraps enter the drain when you use bars of soap until only a small remnant remains. Unfortunately, such a remnant is small enough to enter the drain and large enough to contribute to a clog. To prevent a clogged drain, dispose of the soap in the wastebasket before it becomes a sliver.

KITCHEN

Most kitchen sinks are equipped with a strainer basket to prevent food particles or other items from entering the drain. If your sink has no strainer, buy one at a home center.

If you have a garbage disposal, check the manufacturer's use and care manual to learn what food materials the disposal can handle without becoming clogged. Insert food scraps that it can handle loosely into the drain; tightly packed scraps may clog the machine. Run cold water into the drain while the disposal is operating. The cold water will aid the grinding process and congeal grease so it flushes out of the machine and through the drain. After using the machine, let the cold water run for a minute or two to flush food residue out of the drainpipe. When the disposal is not in use, place the cover over the drain so that stray kitchen items such as tableware do not fall into it, causing a clog or damage to the machine.

Tip Box

Don't use chemical drain cleaners or boiling water in a disposal drain. The chemicals may damage plastic or other parts of the disposal. These cleaners are caustic and may burn skin or eyes if splashed into them. For other drains, wear eye goggles and take care to avoid splashing chemical cleaners when using a drain plunger. Avoid lying under a drain trap during trap removal. Drain cleaner residue in the trap could drip into your face or eyes.

Cleaning Traps

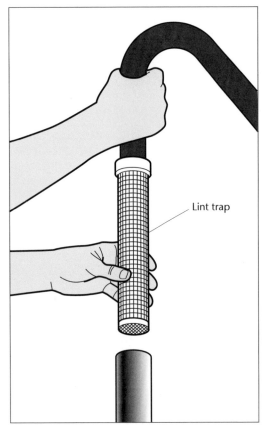
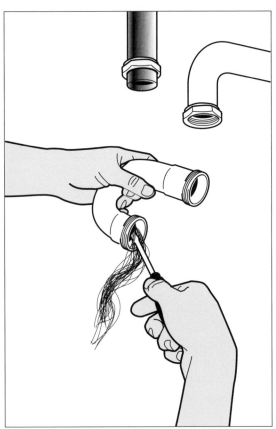

Lint trap

To prevent drain clogs, install a lint trap over the end of your washing machine's water discharge hose.

LAUNDRY

Laundry machines can also become a likely site for drain clogs. Install a nylon mesh lint trap over the clothes washer's discharge hose. This will catch any lint from the clothes so it does not enter the drain. (See "Cleaning Traps" Illustration above.)

The hot and cold water hoses connect the washing machine to the water pipes. Turn the faucets on to do the laundry. After the laundry is done, turn off the shutoff valves on the hot and cold water lines. If left on, the water hoses will perform like a water pipe and will be under constant water pressure. If the water hoses burst when you are away, you will come home to an overloaded drain and a flooded basement.

UNCLOGGING A DRAIN

In spite of all your preventive efforts, you may have a clogged drain. Several solutions are possible, and you should try to solve the problem yourself before calling a plumber. If a fixture clogs, the first step should be to use water pump (channel-type) pliers to loosen the slip nuts on the drain trap. Keep a plastic pail handy to catch the water from the trap. Now inspect the trap: remove any debris and clean the inside of the trap. (See "Cleaning Traps" Illustration above.) If the trap is not clogged, use a

Using a Plunger

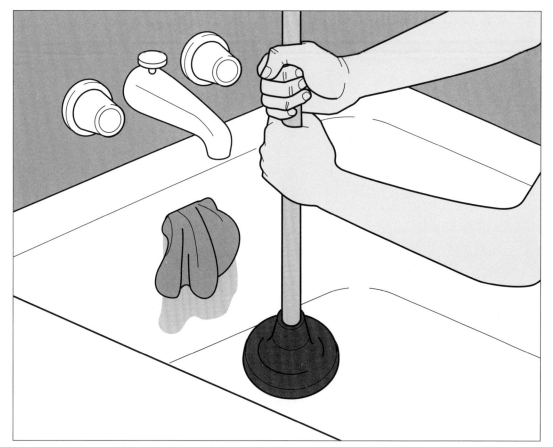

For minor drain clogs, use a plunger to force air into the drain. Cover the fixture's drain hole with the rubber plunger cup, then push the plunger handle rapidly up and down to loosen the clog. If you are using the plunger in a sink or tub that has an overflow, push a wet towel into the overflow to prevent the air from flowing out rather than being forced down the drainpipe.

drain auger to open the branch drainpipe.

An ordinary drain plunger, called a "plumber's friend," is a good choice for clearing minor drain clogs in most drain locations. If you are using the plunger on a fixture that has an overflow drain, such as a sink or a tub, use a towel to plug the overflow opening. This will prevent air from the plunger escaping out the overflow. Wet the edge of the plunger cup and set the plunger cup over the drain. Push down and up rapidly and forcibly on the plunger handle to force air into the drain and expel the clog. It may take repeated attempts to clear it away. (See "Using a Plunger" Illustration above.)

Chemical drain cleaners are widely advertised for clearing clogged drains. However, it is advisable to consider the probable cause of the clog before opting for a chemical remedy. If the clog is in a kitchen sink drain, it is probably the result of food or grease getting into the drain. Chemical cleaners are a reasonable choice for clearing these clogs, but be aware that these chemical cleaners

Closet Auger

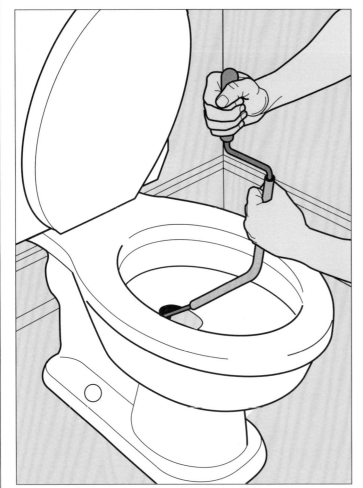

Push the auger into the toilet bowl, and rotate the auger as you feed it deeper into the drain.

it deeper into the drain. A crank handle allows you to rotate the cable as you insert it into the drainpipe. The closet auger is handy for cleaning short runs on branch drains. You can buy a closet auger for home use, or rent a larger auger unit at tool rental stores.

If you have a clogged main drain, you can attempt to remove the clog yourself by renting a power sewer auger. However, this is a messy and sometimes diffi-cult job, so it is advisable to call a professional plumber to clean the main drainpipe. Plumbers charge about $100 for this service. It's also advisable to consider main drain cleaning to be a pre-ventive maintenance job, rather than a last resort emergency measure after the sewer has overflowed. If you have ever dealt with a sewer backup, you will understand the wisdom of having sewer mains cleaned periodically to avoid the mess.

SUMP PUMP

A sump pump is required in any house that has a drain problem below the level at which the main drain exits the house. A sump pump may be used to prevent groundwater from entering the basement, or to serve as a drain for a below-grade laundry room. (See "Sump Pump" Illustration on page

may be caustic. Carefully read and observe the product directions and avoid splash-ing the chemical cleaners into your face or eyes.

Any horizontal drainpipe run must have a cleanout plug so there is access to clean the drain. This plug is usually located at the high end of the drainpipe, that is, where the pipe begins to

slope downward. The cleanout plug offers access to a plumber's snake and is placed so the access is in the direction of the drain flow.

A closet auger is a length of steel cable designed to clean toilet drains. (See "Closet Auger" Illustration above.) Just push the auger into the toilet bowl, and rotate the auger as you feed

Sump Pump

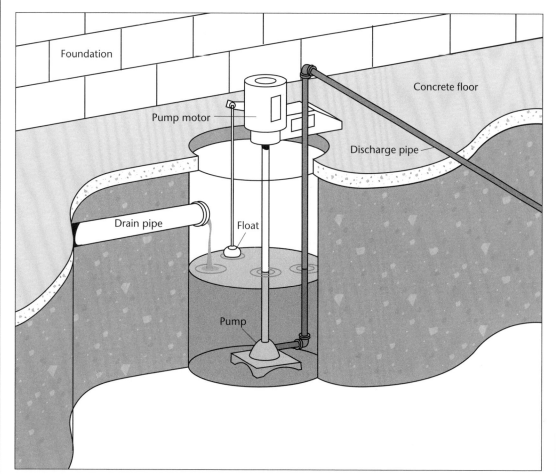

Foundation

Concrete floor

Pump motor

Discharge pipe

Drain pipe

Float

Pump

Typical sump pump installation. If the float ball will not rise, use steel wool to clean the float rod. If this does not solve the problem, unscrew and replace the float ball.

53.) If you have a sump pump in your house, the typical repair problem involves correcting or replacing a faulty float ball. The float ball is attached to a rod that travels through guides. As the float ball rises, it trips the pump switch and turns on the pump. As the water in the sump is pumped out, the water level falls and the float ball falls with it. When the sump is empty, the float rod shuts off the pump until the sump refills with water, and the cycle is repeated. The most common problem is a leaky float ball that fills with water and cannot rise to activate the pump switch. To service the sump pump, first disconnect the power cord. The power cord is plugged into a grounded electrical receptacle at least 6 feet (2 m) above the bottom of the sump. Now check the float ball and rod to be sure the rod is not corroded and travels freely in its guides. If the rod moves freely when you move it by hand, the likely problem is a leaky float ball. The float ball can be unscrewed from the float rod for replacement with a new float ball.

PART II

MAKING REPAIRS

CHAPTER 6

Water Leaks

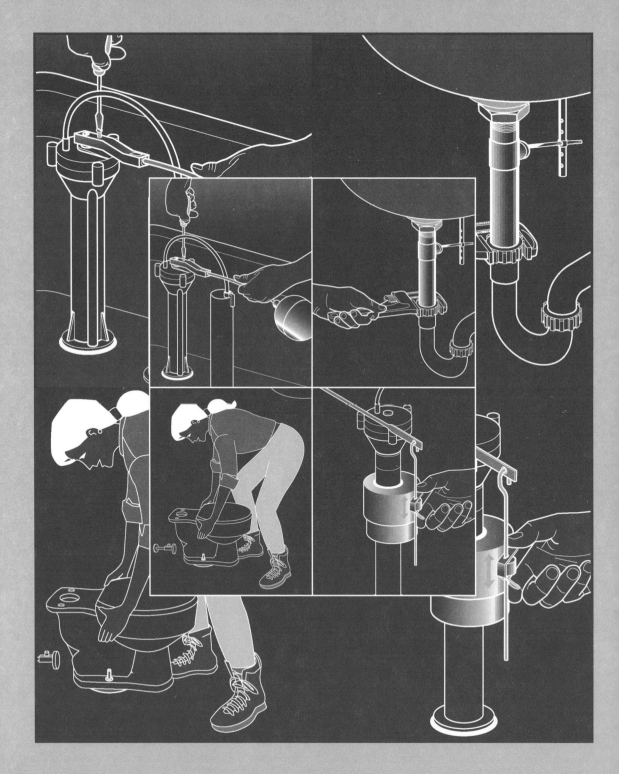

If your water service pipes are leak-free, leaks are most likely to occur in your fixtures or faucets. Repairing these types of leaks is usually a simple job, well within the skills of the home handyperson. In this chapter we will review, room by room, the most common sources of water leaks other than faucets and how to repair them. See Chapter 7 for faucet repair.

To make these repairs, you will need basic tools such as slot and Phillips head screwdrivers, an adjustable wrench, water-pump (channel-type) pliers and Teflon plumber's tape. Wrap the Teflon tape around the threads of each fitting to ensure that its joints are leak-proof. You will also need towels and a plastic pail to catch the water when you open fitting joints.

BATHROOM

The toilet is the source of most bathroom leaks. A common place for a leak to occur is in the tank, where a faulty float, or the seal between the flapper or ball and the seat, causes the toilet to run continuously. If the toilet ball cock does not shut off, first check the float ball (if your toilet has one) to be sure the ball has not

Leaky Tank Ball

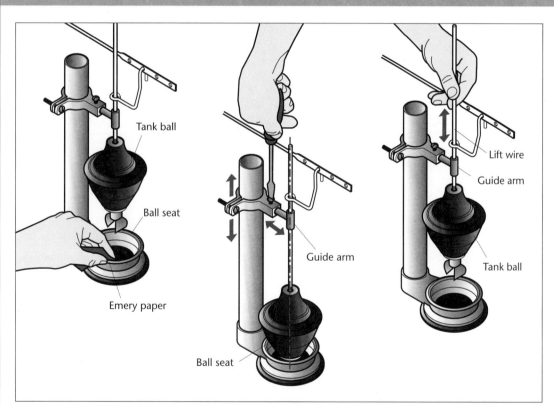

Shut off the water valve and flush the toilet. Lift the flapper or ball and check for mineral deposits on the ball seat. If the seat feels rough, use emery cloth to smooth the seat and the bottom of the ball or flapper.

Adjust the guide arm so it is exactly aligned over the ball seat. Check the length of the chain so the chain is extended when the ball or flapper is closed.

Pull up the lift wire to be sure it moves smoothly through the wire guides. Be sure the float ball does not touch the water tank.

leaked and filled with water. If the float ball contains water, it will not float upward and shut off the ball cock valve. Replace the old float ball with a new one, or replace the entire assembly with a new float-cup valve. (See "Float-Cup Valve" Illustration on page 59.)

When the toilet flush lever is depressed, the flapper or ball lifts to permit water to run out of the tank and flush away the contents of the toilet bowl. After the flush is completed, the flapper or ball falls back to seal the water outlet drain or ball seat from the tank to the bowl. If the seal is complete, the water tank will refill. If not, water will leak into the toilet bowl and the water intake valve or ball cock will not shut off. You will hear water running intermittently or continuously when the toilet has not been flushed.

The first step in remedying this problem involves checking the flapper or ball and the rim of the toilet tank drain or ball seat to see if there is any sediment present. (See "Leaky Tank Ball" Illustrations on page 57.) If there is, clean the bottom of the flapper ball and the rim of the ball seat with fine emery cloth or steel wool.

Now flush the toilet. If the flapper or ball seals the ball seat, the water valve or

Adjusting a Float Arm

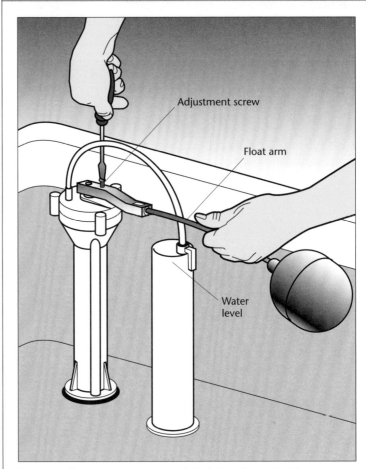

Turn the adjustment screw on top of the ball cock to adjust the water level. You can also bend the float arm slightly to adjust the water level in the tank.

ball cock will shut off when the toilet tank is refilled. If this does not do the job, you must replace the flapper or tank ball.

If you have an old-style ball cock, with a float arm and float ball, you can simplify repairs by replacing the old mechanism with a new float-cup valve assembly. To install this new assembly, shut off the water valve on

the riser tube beneath the toilet tank. Flush the toilet to remove the water. When the tank has drained, use a towel to wipe the inside of the tank dry.

Use an adjustable wrench to unscrew the connecting nut on the top of the riser tube. Now use channel-type pliers to unscrew the retaining nut that holds the ball cock in place. When the nut

Float-Cup Valve

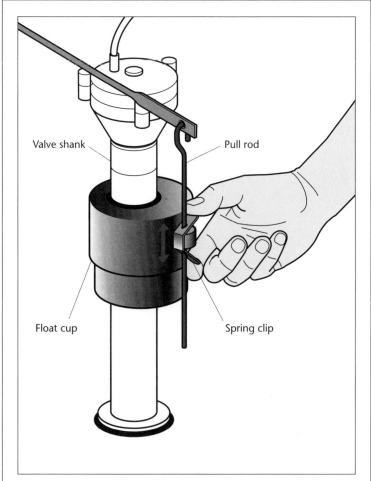

Valve shank

Pull rod

Float cup

Spring clip

If you have a float cup, pinch the clip and raise or lower the cup on the pull rod to adjust the water level.

Valve" Illustration at left.) Check for leaks at the joint between the riser tube and the new float assembly.

WAX RING

Condensation occurs on the toilet tank or pipes when cold water fills the warm tank. If you see water on the floor around the base of the toilet stool, check first for condensation on the tank. If there is evidence of it, you can buy a terrycloth cover that will insulate the tank and absorb any condensation.

If there is no condensation present, the problem may be a faulty wax ring. The wax ring is a doughnut made of wax that seals the joint between the toilet stool and the drainpipe the toilet is connected to. Replacing the wax ring involves several steps but is not difficult.

First, shut off the water valve beneath the toilet

is removed, lift the old ball cock out of the tank.

Now install the new float-valve. Push the threaded end of the float-cup valve through the hole in the bottom of the tank and secure it in place with the retaining nut. Reattach the riser tube to the nipple on the cup assembly, and turn until watertight. Insert the water refill tube into the overflow tube.

Turn the water back on and let the toilet tank fill until the water level is approximately 1 inch (2.5 cm) below the top of the overflow tube. To adjust the water level in the tank, pinch the spring clip on the pull rod and slide the float cup on the assembly shank up (to raise the water level) or down (to lower the water level). (Refer to "Float-Cup

Tip Box

Never use caulk to seal the joint between the floor and the toilet stool base. If there is a leak around the wax ring, the caulk will trap the water between the stool and the floor, causing water damage to the floor. Leave the stool and floor joint unsealed so leaking water will be visible.

Replacing a Toilet Wax Ring

Clean away any residual wax from the old wax ring, both from the floor and from the bottom of the bowl.

Place the new wax ring over the hole in the bottom of the toilet bowl, pressing it firmly in place.

Position it over the hole, using the retaining bolts as a guide. When the bowl is in place, press it down and rock it from side to side to ensure a good seal between the bowl and the floor.

tank. Now flush the toilet to remove the water from the tank. Use a towel to wipe the inside of the toilet tank dry.

Next, remove the toilet tank from the bowl. The tank may be attached to the wall with screws, or may be screwed to the toilet bowl itself. Look inside the tank: if the tank is attached to the bowl, there will be two retaining screws with rubber washers in the bottom of the tank. Remove these screws and lift the toilet tank away, being careful not to break it.

With the tank removed, pull off the caps that conceal the bolts over the base of the toilet bowl. There will be either two or four of these caps and bolts. Use a wrench to remove the nuts from the toilet bolts. When the nuts are removed, you can gently lift away the toilet bowl, taking care not to break it. Clean away any residual wax from the old wax ring, both from the floor and from the bottom of the bowl.

Place the new wax ring over the hole in the bottom of the toilet bowl, pressing it firmly in place. Turn the bowl over and position it over the hole, using the retaining bolts as a guide. When the bowl is in place, press it down and rock it from side to side to ensure a good seal between the bowl and the floor.

Replace and tighten the nuts on the toilet retaining

Removing a Fixture Trap

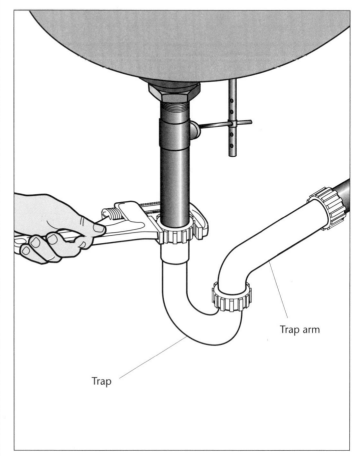

Trap arm

Trap

Use channel-type pliers or a pipe wrench to loosen the slip nuts, then remove the drain trap.

bolts, turning them hand-tight before using a wrench for the final tightening. Slightly tighten the nuts in turn: if you try to completely tighten one bolt at a time, you will put uneven pressure on the bowl and crack it.

Replace the bolt covers or caps to conceal the bolts from view. If necessary, use plumber's putty in the bolt covers to hold them in place. Flush the toilet and check

for leaks at the base of the stool.

HOT WATER HEATER

If you see water around the base of your water heater, first check the water to see if it is rusty. If it is not, check to see if the water pipes at the top of the heater are leaking or if there is condensation on the pipe. If there is

Replacing an Old Trap

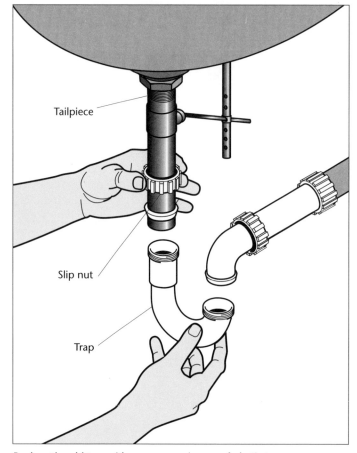

Tailpiece

Slip nut

Trap

Replace the old trap with a new corrosion-proof plastic trap.

as water pipes. If you forget to turn the valves off after doing laundry, the hoses may leak while you are away and flood the laundry room.

If you see water on the floor around your washer, check the water hoses for leaks. Replacement hoses are inexpensive and easy to change: just be sure to connect the cold water valve to the cold water side of the washer.

TRAPS

We have discussed trap leaks elsewhere in this text, but the solution to this problem bears repeating. If a metal fixture trap is leaking, it has probably rusted through. Examine the trap at the bottom to see if there is a rust hole. If none is visible, try tightening the slip nuts at each end of the trap.

If there is a rust hole, you will need to replace the fixture trap. Before doing so, you will first need to set a plastic pail under the trap to catch the water. Now use channel-type pliers to loosen the slip nuts on each end of the trap. (Refer to "Removing a Fixture Trap" Illustration on page 61.) Pull the trap away and take it along to the hardware store to be sure you get a replacement trap of identical size and shape. Once you have a replacement, position the trap so the top end is aligned

condensation, wrap the pipes with pipe insulation. Now check the drain valve at the bottom of the heater to be sure it is turned off. If the pipes are neither leaking nor forming condensation and the drain valve is closed, the water heater has developed a leak and must be replaced. See Chapter 8: Installing Appliances, for directions on replacing the water heater.

CLOTHES WASHER

There are shutoff valves on both the hot and cold water lines to your clothes washer. These valves should be turned on when doing the laundry and should be turned off when the washer is not in use. When the valves are turned on, the rubber hoses are under the same water pressure as the water pipes, and the hoses are not designed to function with the tailpiece from the sink and the lower end is aligned with the trap arm from the drain. Tighten the slip nuts by hand; then use the channel-type pliers to turn them until they are watertight. (See "Replacing an Old Trap" Illustration on page 62.)

CHAPTER 7

Faucet Repairs

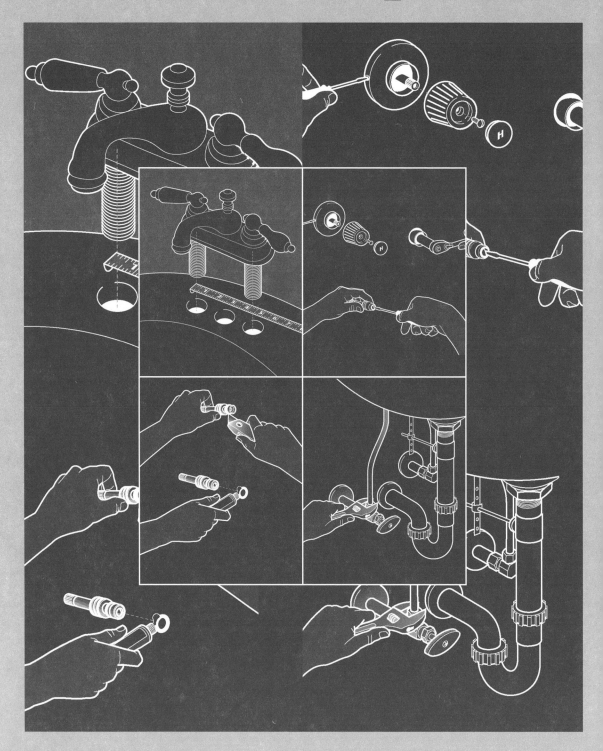

There are several types of faucets, so you must know which type you have before buying repair parts. In older houses the faucets were all compression faucets (also known as stem and seat faucets). In modern houses most faucets are either ball-type or cartridge faucets. Cartridge faucets come in two types: either disc-type or sleeve-type.

To simplify faucet repair, keep the manufacturer's install and repair instructions when you install a new faucet. If you are having a new house built, ask your builder to note the faucet manufacturer and model, and file this information—with warranties for all the appliances and equipment in your new home. This file will prove an invaluable reference if you find you need to repair or replace something in your new house.

Before you start to repair a faucet, you should be aware of the common characteristics shared by faucets. With any faucet, you must remove the faucet handle to begin repairs. To remove the handle on a ball-type faucet, you will need a hex (Allen) wrench. To take off the handle on all other faucets, you will need a screwdriver with either a slot or a Phillips tip. The faucet handle screw is exposed on older compression-type faucets. On newer models, an index cap or cover conceals the handle screw. If you have an index cap on your faucet, use a small screwdriver to gently pry it off and gain access to the handle screw. Be careful not to scratch the cap when removing it.

If you know the model and make of your faucet, you can use this information to buy the proper replacement parts. If you don't have this information, take the faucet apart and take the old parts to your hardware dealer to be sure the replacement parts fit the old faucet.

Service valves permit you to turn off the water supply to one fixture while making repairs. Turn off the service valves under the faucet you are repairing. If you have an older house, there may not be service valves to individual fixtures. In this case, you will have to shut down the water system of the entire house via the main shutoff valve. Once you've turned off the water, open the faucet to drain any remaining water and relieve the pressure on the water pipe.

Compression faucets have neoprene washers and O-rings. You can buy an assortment pack containing a variety of washers and O-rings so you have a supply on hand. Use heatproof plumbing grease on all washers and O-rings.

If you have separate hot and cold faucets, perhaps in a laundry tub or an older sink, the hot water faucet will usually begin to leak first. Be aware that the cold water faucet will soon leak too, so service both faucets at the same time to avoid having to repeat the procedure in a few weeks.

COMPRESSION FAUCETS

Repairing a compression faucet is a fairly simple job. You will need a screwdriver, an adjustable wrench, a utility (razor) knife, washers and O-rings or packing string if you have an older model.

Shut off the water supply valve and open the faucet to relieve the water pressure and drain any remaining water.

Tip Box

Leaking faucets are a common home-plumbing problem. Left untended, a leaking faucet wastes water, stains fixtures and disturbs the homeowner's sleep. Depending on nature and location of the leak, it can also cause water damage to the surrounding area. A common mistake homeowners make is to crank hard on the faucet handle to try to stop the leak. This action can distort the faucet body and ruin the faucet. If a faucet leaks, repair it.

Compression Faucet

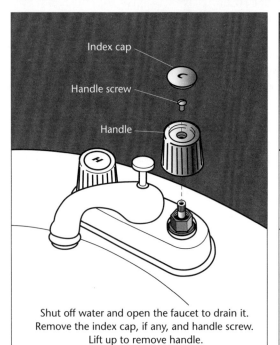

Shut off water and open the faucet to drain it. Remove the index cap, if any, and handle screw. Lift up to remove handle.

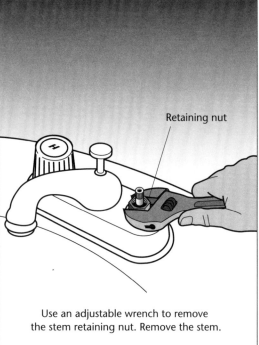

Use an adjustable wrench to remove the stem retaining nut. Remove the stem.

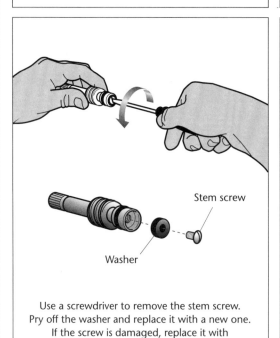

Use a screwdriver to remove the stem screw. Pry off the washer and replace it with a new one. If the screw is damaged, replace it with one from the repair kit.

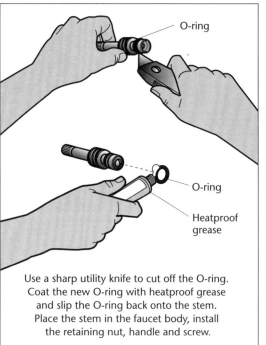

Use a sharp utility knife to cut off the O-ring. Coat the new O-ring with heatproof grease and slip the O-ring back onto the stem. Place the stem in the faucet body, install the retaining nut, handle and screw.

Wall-Mounted Compression Faucet

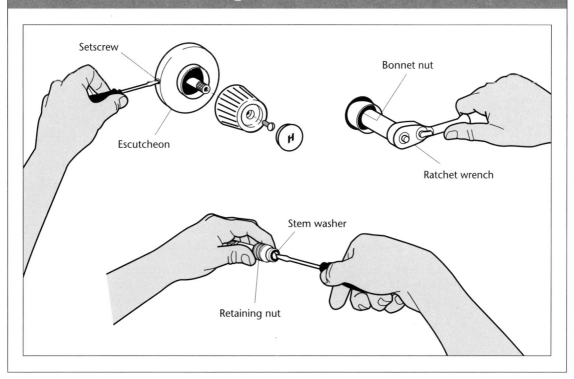

If there is an index cap on the top of the handle, carefully pry off the cap. Use a screwdriver to remove the handle screw. Use the adjustable wrench, or channel-type pliers, to loosen the retaining nut, then turn it out by hand. Remove the stem and use a screwdriver to remove the washer from the stem. Inspect the brass washer screw: if the screw is worn or damaged replace it with a new one. Install the new neoprene washer and the screw. Now use the utility knife to cut away the O-rings. (Tip: always use a sharp blade in a utility knife. Most people are injured by dull knives, rather than by

sharp ones, because they force the dull knife.) Roll new O-rings onto the stem. Coat the washers and O-rings with heatproof grease.

Position the stem in the faucet body and turn the retaining nut until it is hand-tight. Use the adjustable wrench to finish tightening the retaining nut. Place the faucet handle back on the stem and replace the screw. Test the faucet to be sure it is leak-free. (See "Compression Faucet" Illustration on page 66.)

Many bathtub faucets are compression or stem-and-seat types. The retaining nut and stem are recessed below the level of the bath tile, so

they cannot be reached with an ordinary wrench. To repair the bathtub faucet, first remove the index cap and handle as above, then loosen the setscrew to remove the escutcheon. You must use a ratchet wrench and deep socket to remove and install the stem retaining nut. Aside from this step, the repair process is the same as for other compression faucets. (See "Wall-Mounted Compression Faucet" Ilustration above.)

BALL-TYPE FAUCET

To repair a ball-type faucet, you will need a screwdriver,

Ball-Type Faucet

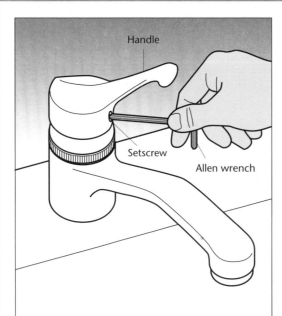

Shut off water and open faucet to drain water. Use an Allen wrench to remove the handle setscrew, then lift off the handle.

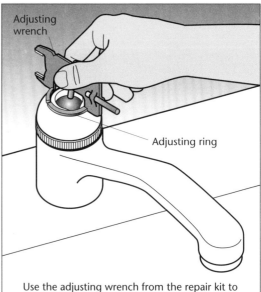

Use the adjusting wrench from the repair kit to tighten the adjusting ring. Replace the handle and turn on the water. If the leak is fixed, stop here. If not, turn off the water and remove the handle.

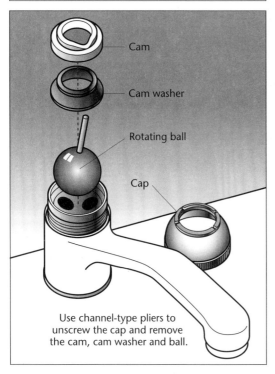

Use channel-type pliers to unscrew the cap and remove the cam, cam washer and ball.

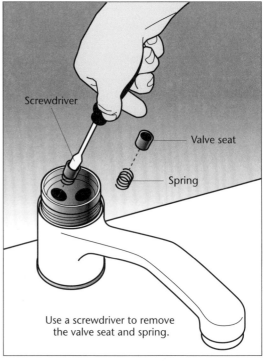

Use a screwdriver to remove the valve seat and spring.

Ball-Type Faucet

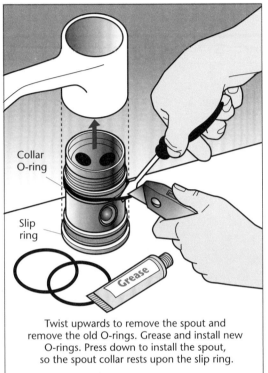

Collar O-ring

Slip ring

Grease

Twist upwards to remove the spout and remove the old O-rings. Grease and install new O-rings. Press down to install the spout, so the spout collar rests upon the slip ring.

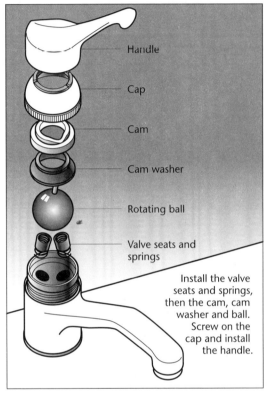

Handle

Cap

Cam

Cam washer

Rotating ball

Valve seats and springs

Install the valve seats and springs, then the cam, cam washer and ball. Screw on the cap and install the handle.

channel-type pliers, an Allen wrench, a utility knife, and a repair kit or kits.

Shut off the water supply valve and open the faucet to relieve the water pressure and drain any remaining water.

Use the Allen wrench to loosen the handle setscrew. Use the wrench that is included in the repair kit to tighten the adjusting ring, and replace the handle. Turn on the water: tightening the adjusting ring may have stopped the leak.

If the faucet still leaks, turn the water off and remove the handle. Use the

channel-type pliers to remove the faucet cap, and pull out the cam, cam washer and ball. Use the screwdriver to remove the springs and valve seats. To remove the faucet spout, twist it upward. Remove and replace the O-rings. Push down on the spout until the spout collar seals against the slip ring.

Replace the springs, valve seats, cam, cam washer and ball. Replace the faucet handle and tighten the setscrew. Turn on the water and check for leaks (See "Ball-Type Faucet" Illustration on pages 68–69.)

SLEEVE-TYPE CARTRIDGE FAUCET

To repair a sleeve-type cartridge faucet, you will need both channel-type and needle-nose pliers, a utility knife, a screwdriver, a replacement cartridge and O-rings.

Shut off the water supply valve and open the faucet to relieve the water pressure and drain any remaining water.

Remove the index cap and handle screw. Holding the lever up, lift the handle

Sleeve-Type Cartridge Faucet

Shut off the water and open the faucet. Pry off the index cap and remove the handle screw. Holding the handle upright, pull the handle off.

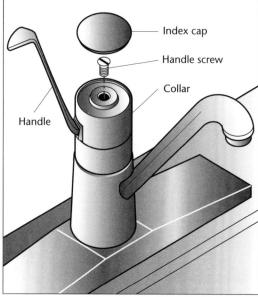

Index cap

Handle screw

Collar

Handle

Use the channel-type pliers to remove the retaining nut. Also remove the grooved collar under the nut, if there is one.

Retaining nut

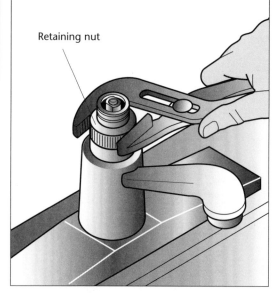

Use needle-nose pliers to pull the retaining clip from the top of the cartridge.

Retaining clip

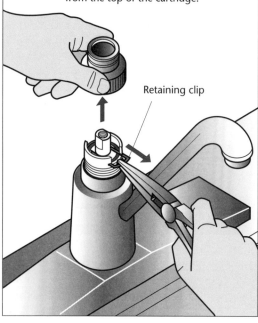

Use the channel-type pliers to pull the old cartridge from the faucet. Install the new cartridge and retaining clip.

New sleeve-type cartridge

O-ring

Seals

Old cartridge

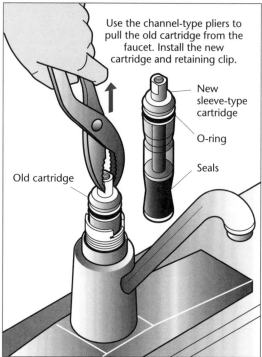

Sleeve-Type Cartridge Faucet

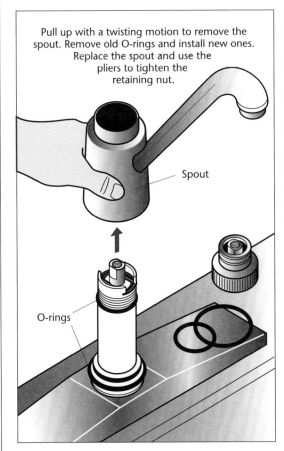

Pull up with a twisting motion to remove the spout. Remove old O-rings and install new ones. Replace the spout and use the pliers to tighten the retaining nut.

Spout

O-rings

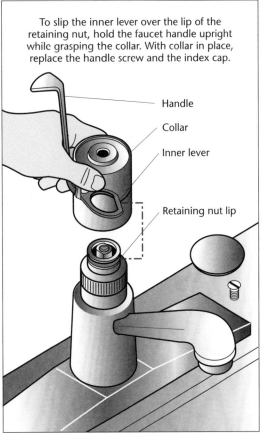

To slip the inner lever over the lip of the retaining nut, hold the faucet handle upright while grasping the collar. With collar in place, replace the handle screw and the index cap.

Handle

Collar

Inner lever

Retaining nut lip

off the faucet. Use the channel-type pliers to remove the retaining nut, and remove the collar if any. Use the needle-nose pliers to remove the retaining clip; then use the channel-type pliers to pull the cartridge upwards. Using a twisting motion, pull up to remove the spout. Use the utility knife to cut off the old spout O-rings. Coat the O-rings with heatproof grease and replace the spout and the retaining nut. Hold the faucet lever fully upwards

and slip the collar over the retaining nut. Replace the handle screw and press in the index cap. (See "Sleeve-Type Cartridge Faucet" Illustration above.)

DISC-TYPE CARTRIDGE FAUCET

To repair a disc-type faucet, you will need a screwdriver, an Allen wrench and a new cartridge.

Shut off the water supply

valve and open the faucet to relieve the water pressure and drain any remaining water.

Pry off the index cap and remove the handle screw. Pull up to remove the faucet handle. Loosen the setscrew with the Allen wrench and remove the handle insert. Now unscrew the dome cap and remove the cartridge mounting screws. Lift out the old cartridge, and insert the new one into the faucet. Install the mounting screws

Disk-Type Cartridge Faucet

Shut off the water and open the faucet. Use a screwdriver to pry off the index cap; then remove the handle screw and faucet handle.

Handle screw

Index cap

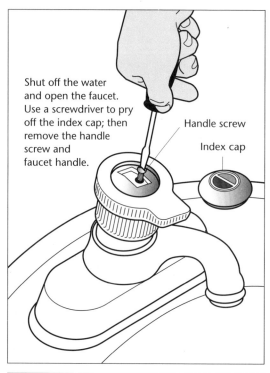

Loosen the setscrew with an Allen wrench and remove the handle insert. Unscrew and set aside the dome cap.

Hex (Allen) wrench

Setscrew

Handle insert

Dome cap

Remove the cartridge mounting screws and pull out the cartridge.

Replace the new cartridge and tighten the mounting screws. Screw on the dome cap and reassemble the handle insert, handle, handle screw and index cap.

Index cap

Handle screw

Handle

Dome cap

Setscrew

Handle insert

Mounting screws

Cartridge

and the dome cap, then the handle insert and handle. Replace the handle screw and the index cap. (See "Disk-Type Cartridge Faucet" Illustration on page 72.) Turn on the water and check for leaks.

REPLACING A FAUCET

You may decide to replace a faucet that has had a lot of wear and tear or you may want to replace an older model with a newer one. For example, you can replace an old compression faucet with a new single-lever model that will provide years of trouble-free operation. The single lever controls both water volume and temperature. For those who have difficulty gripping the handle it can be operated by a push of the hand or arm.

If you decide to replace a faucet, shut off the water supply valve and open the faucet to relieve the water pressure and drain any remaining water.

Disconnect the riser tubes from the faucet tailpieces. Remove the faucet from the sink. The faucet is held in place by retaining nuts on the tailpieces. Because the basin of a sink drops down from the sink rim, you cannot reach the faucet retaining nuts with an ordinary wrench. Buy or rent a basin wrench, a long-han-

Faucet Replacement

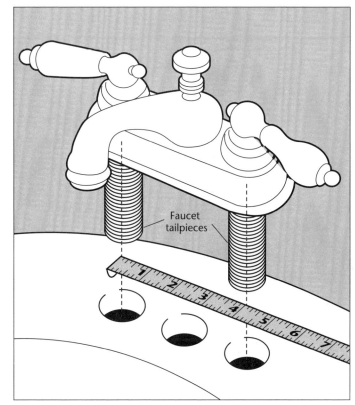

Faucet tailpieces

If the faucet is worn, replacement may be the solution. Measure the distance between the faucet tailpieces, or take the old faucet to the hardware store to make sure the new faucet will fit the sink.

dled wrench that lets you reach past the sink basin to remove the retaining nuts.

With the retaining nuts removed, lift away the old faucet. Measure the distance from center to center of the old tailpieces. The replacement faucet must have matching tailpieces in order to fit the faucet holes in the sink. When ready to install the new faucet, position the unit tailpieces in the sink holes and tighten the retaining nuts hand-tight. Then

use the basin wrench to further tighten them. Reattach the riser tubes to the faucet tailpieces and turn on the water. Check for leaks in the riser tubes and tailpieces and tighten if necessary. (See "Faucet Replacement" Illustration above.)

FIXTURE SHUTOFF VALVES

Fixture shutoff valves permit you to shut off the water while repairing a single fix-

Installing Fixture Shutoff Valves

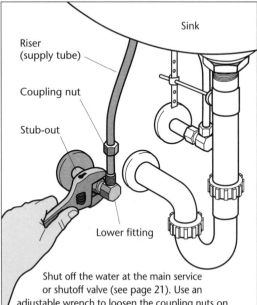

Riser
(supply tube)

Coupling nut

Stub-out

Sink

Lower fitting

Shut off the water at the main service
or shutoff valve (see page 21). Use an
adjustable wrench to loosen the coupling nuts on
the lower fitting and the stub-out. Pull the fitting
off the supply tube and the stub-out. Use a hacksaw
to cut the compression ring off the stub-out.

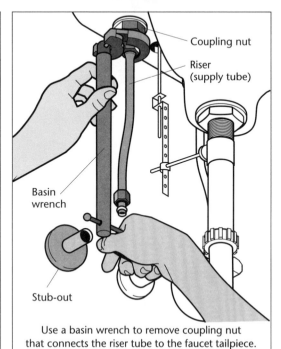

Coupling nut

Riser
(supply tube)

Basin
wrench

Stub-out

Use a basin wrench to remove coupling nut
that connects the riser tube to the faucet tailpiece.

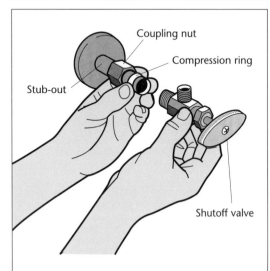

Coupling nut

Compression ring

Stub-out

Shutoff valve

To install the shutoff valve, slide a new coupling nut
and compression ring onto the stub-out. Install the
shutoff valve so the outlet is on top. Use one wrench
to hold the valve and prevent it from turning while
you tighten the coupling nut with the other wrench.

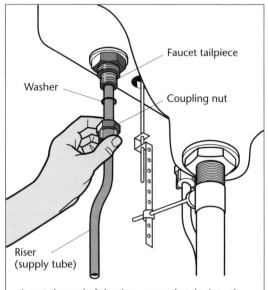

Faucet tailpiece

Washer

Coupling nut

Riser
(supply tube)

Insert the end of the riser or supply tube into the
faucet tailpiece. Slide a washer and the coupling
nut up the riser and hand-tighten the nut
on the tailpiece. Use the basin wrench
for final tightening.

Installing Fixture Shutoff Valves

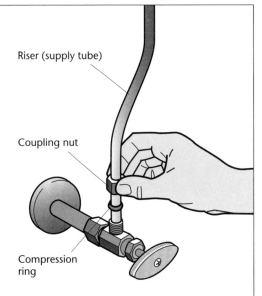

At the shutoff valve, slide the coupling nut and compression ring onto the end of the riser or supply tube. Push the compression ring tightly against the faucet; then hand-tighten the coupling nut on the valve.

Riser (supply tube)

Coupling nut

Compression ring

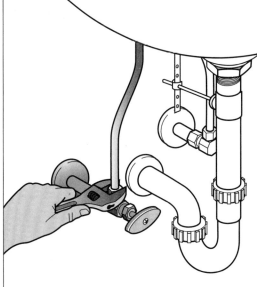

Use the adjustable wrench to tighten the coupling nut; then turn the water on and test for leaks. If there is a leak, tighten the nut an extra half-turn and test again.

ture, without shutting down the water supply to the entire house. In older houses there is a shutoff valve only on the toilet stool, the fixture that is most used and most repaired. But to make repairs at sinks and laundry tubs, you must shut off the water at the main shutoff valve (see page 23). This means you cannot use any other fixtures in the house while a single fixture is being repaired.

In newer houses fixture shutoff valves are commonly installed on all fixtures. This is a great convenience. If you have to repair or replace faucets, consider installing a fixture shutoff valve on the fixture being repaired, or make a major upgrade and install these valves on all fixtures.

When you install a fixture shutoff valve, you must also replace the riser or supply tubes to the fixture faucets. (See "Installing Fixture Shutoff Valves" Illustration on pages 74–75.)

PART III

MAKING IMPROVEMENTS

CHAPTER 8

Installing Appliances

Installing certain appliances is within the skills of the home handyperson, but be aware that local codes may require that you obtain permits and inspection from the local building department. The call for permits is especially critical when you are working with utilities such as natural gas or electricity. Because mistakes you may make with these utilities can be dangerous or even fatal, you should consider local codes and your own level of expertise before undertaking installations that involve disturbing or adding onto electrical or natural gas lines.

WASHING MACHINE

The simplest appliance to install is a washing machine. Turn off both the hot and cold water supply valves and unplug the power cord to the washing machine.

When you unpack the new clothes washer, first read the installation instructions. Because the washer tub is assembled on flexible mounts that permit it to gyrate during the spin cycle, a temporary shipping brace is used to secure the tub during shipping and handling. The temporary brace must be removed before using the clothes washer. It may be mounted on the bottom of the washer, so you must lay

the washer on its back to remove the brace. Once you have removed it, move the washer to its proper position. Use an adjustable wrench to adjust the legs to level the machine. This is important for the operation and life of the machine. Attach the water hoses to the shutoff valves, being careful to attach the cold water hose to the cold side, and the hot water hose to the washer's hot side. The hot and cold water inlets will be marked on the back of the machine.

The electrical outlet for appliances that use both electricity and water should be equipped with a ground fault circuit interrupter (GFCI). A GFCI monitors the outlet to ensure that the electrical current flow through the neutral (white) wire is the same as the current on the hot (black) wire. Any drop in current flow on the neutral wire indicates the current is going to ground, and the GFCI instantly shuts off the receptacle to prevent dangerous or fatal electrical shock to the operator. U.S. electrical codes require that all electrical receptacles in the kitchen, bath, laundry or exterior outlets be equipped with GFCI protection.

If the electrical receptacle at your washing machine does not have a GFCI, install one. A GFCI is inexpensive and easy to install. Directions for installing it

are included with the unit. Be sure to shut off the electrical power to the circuit before you begin. To be sure the circuit is dead, check the receptacle with a circuit tester before proceeding.

INSTALLING A WATER HEATER

If you hear a rumbling noise when the burner of your water heater burner fires up, or see water—perhaps rusty water—beneath the unit, it's time to replace the heater.

The instructions that follow refer to installing a water heater that is fired by natural gas and vents upward through a vertical vent, not through a wall. Some new water heaters have a forced draft feature for better fuel efficiency. The process for installing this type of heater is basically the same as for a vertical vent heater, except that the vent pipes are plastic and can be vented through a wall. The installation instructions provided here are general, but if they vary from the instructions that come with your water heater, follow the manufacturer's instructions. (See "Water Heater Installation" Illustration on page 80)

Also, remember that water heater installation problems are about leaks in either the gas or water pipes. Always use Teflon plumber's tape on the threads of all

Water Heater Installation

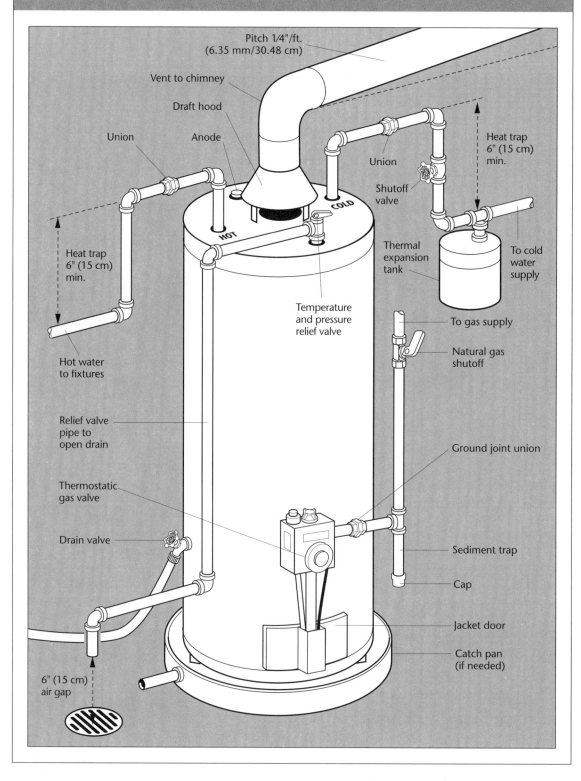

Pitch 1/4"/ft.
(6.35 mm/30.48 cm)

Vent to chimney

Draft hood

Union

Anode

Union

Shutoff valve

Heat trap 6" (15 cm) min.

To cold water supply

HOT

COLD

Thermal expansion tank

Heat trap 6" (15 cm) min.

Temperature and pressure relief valve

To gas supply

Natural gas shutoff

Hot water to fixtures

Relief valve pipe to open drain

Ground joint union

Thermostatic gas valve

Drain valve

Sediment trap

Cap

Jacket door

Catch pan (if needed)

6" (15 cm) air gap

Thermostat

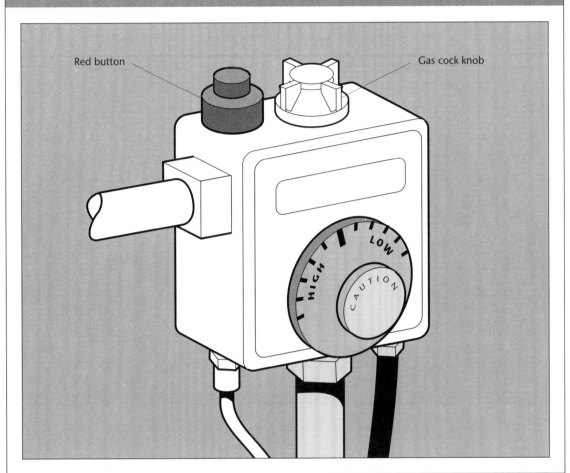

Red button

Gas cock knob

HIGH

LOW

CAUTION

water pipe fittings. Use pipe joint compound on the threads of gas pipe fittings.

To install a gas water heater with copper water pipes, you will need an adjustable wrench, a pipe wrench, adjustable or engineer's pliers, a tubing cutter or hacksaw, a soldering torch, flux and solder, and a Phillips head screwdriver.

Buy a water heater installation kit containing 12-inch (30.48 cm) flexible connectors, a roll of Teflon plumber's tape and a few pipe fittings, such as pipe nipples, which you may need for installing the connectors to the water heater.

Note that you can also use plastic chlorinated polyvinyl chloride (CPVC) pipes and fittings to install the water heater. Using plastic pipe and fittings will eliminate the need to use an

Temperature and Burn Times

120°F (49°C)	more than 5 minutes
125°F (52°C)	1½ to 2 minutes
130°F (55°C)	30 seconds
135°F (57°C)	10 seconds
140°F (60°C)	Less than 5 seconds
155°F (68°C)	1 second

Temperature and Pressure

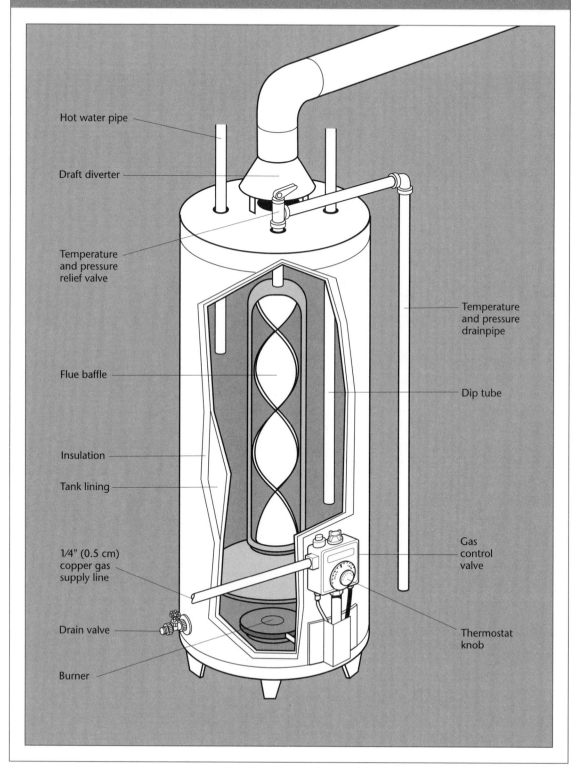

Hot water pipe

Draft diverter

Temperature and pressure relief valve

Flue baffle

Insulation

Tank lining

1/4" (0.5 cm) copper gas supply line

Drain valve

Burner

Temperature and pressure drainpipe

Dip tube

Gas control valve

Thermostat knob

How a Water Heater Works

Cold water enters the top of the heater and flows down the dip tube to the burner at the bottom of the heater. When the water temperature falls, the thermostat calls for heat. The burner ignites and heats the cold water. When a hot water tap is opened, hot water exits the tank. While the burner is lit, hot air and exhaust gases rise up the chimney flue. Hanging in the flue is a twisted steel part called a flue baffle. The flue baffle absorbs and radiates some of the flue heat to help heat the water in the upper portion of the tank.

Also inside the tank is a magnesium rod called an anode. The magnesium anode attracts electrolytes that would otherwise attack and destroy the water heater walls. The anode is eventually consumed by the electrolytes in the water, and can be replaced.

Depending on your water quality, there may be sediment in the water supply. If your water contains sediment, it will settle at the bottom of the heater tank. A valve at the bottom of the tank allows you to drain sediment out. Periodically place a plastic pail under the heater valve and draw off a gallon or so of water. Check it for sediment: if you see any, drain water from the tank more often. You can also install a filter on your water system to improve water quality. Industry estimates place the life of a gas water heater between ten and twelve years.

Either on or within 6 inches (15 cm) of the heater top, there is a valve called the temperature and pressure (T&P) relief valve. This is a safety valve that protects against a possible explosion if the heater malfunctions. If the tank pressure rises above 125 psi, or if the temperature rises above 210 degrees Fahrenheit (99°C), the T&P valve will open and direct the hot water or steam down to floor level. The T&P drainpipe should terminate near a floor drain. (See "Temperature and Pressure" Illustration on page 82.)

water heater as opposed to a natural gas one, get a professional to install it. Because you will be working with both 240-volt electrical power and water—which can be a lethal combination— installing an electrical water heater is beyond the skills of the average homeowner.

Do not forget that in most locales, any project that disturbs or adds onto utilities, such as natural gas and electricity, requires a permit. The finished job must also be inspected.

If you must replace the water heater, consider your hot water requirements. If your family often runs out of hot water, consider installing a larger water heater.

If you undertake to install a gas water heater yourself, be on the alert for gas leaks. If you smell gas during any stage of the installation, shut off the gas valve and vacate the house. If gas is present, do *not* turn on any electrical switch, light a match or even use a telephone, which might cause a spark. Go to an outside phone and call the utility company or fire department and ask them to clear the house of gas fumes.

There is the potential to save a considerable sum of money by installing your own water heater. For a recent installation, a utility company bid $800 U.S., and a plumbing contractor bid $500 plus for a water heater

open flame for soldering. The plastic pipe can be joined directly to the cutoff copper water pipes.

The water pipes to your water heater are ¾ inch in diameter (19.05 mm). To avoid having to make several trips for various pipe fittings, plan what you will need

before taking the piping apart. Draw a diagram of the installation you will make, along with measurements for pipe length. Take the diagram along to the hardware store so a salesperson can help you select the fittings you will require.

If you have an electric

Replacing a Garbage Disposal

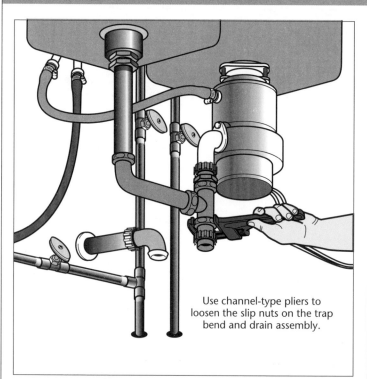

Use channel-type pliers to loosen the slip nuts on the trap bend and drain assembly.

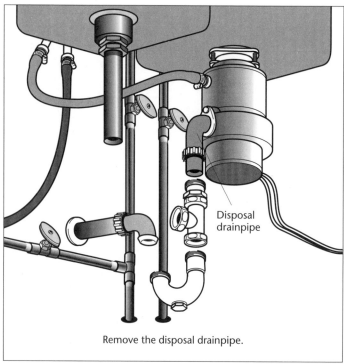

Disposal drainpipe

Remove the disposal drainpipe.

with installation. The same heater, together with 12-inch (30.48 cm) flexible heater connections and assorted fittings, cost $200 at a home center.

Installing a Hot Water Heater

The first step in the installation process involves shutting off both the water valve on the cold water line and the gas valve to the heater you will be replacing. Let the water in the heater cool before starting the project. Open the hot water faucet nearest to the heater, usually by the laundry tub, to drain out as much water as possible. Attach a water hose to the faucet at the bottom of the water heater. Position the open end of the hose in or near a floor drain. With the water and gas supply valves shut off, open the temperature and pressure relief valve located on the top or side of the water heater. (See "Temperature and Pressure" Illustration on page 82.) This open valve will allow air to enter the heater tank so you can completely empty the water from the heater.

Use an open-end or adjustable wrench to loosen the fitting that connects the gas line to the gas control valve. When the gas supply line is disconnected from the heater, remove the screw from the clip that secures the

Removing the Old Disposal Unit

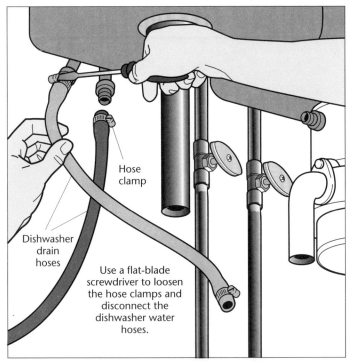

Hose clamp

Dishwasher drain hoses

Use a flat-blade screwdriver to loosen the hose clamps and disconnect the dishwasher water hoses.

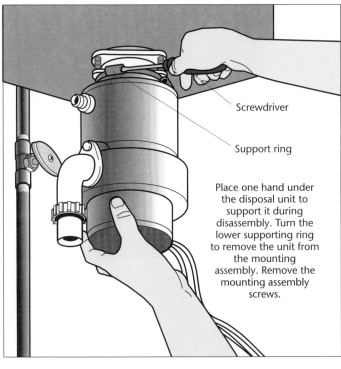

Screwdriver

Support ring

Place one hand under the disposal unit to support it during disassembly. Turn the lower supporting ring to remove the unit from the mounting assembly. Remove the mounting assembly screws.

gas line to the top of the heater. Smell the fitting to see if there is a gas leak. If there is no gas odor, proceed with the project. (As the project progresses, frequently test for gas odors around the line and the gas control valve. And of course never smoke while working on a gas appliance.)

Next, disconnect the water pipes from the water heater. Look for unions or fittings that permit you to disconnect the water lines with a wrench. If there are no unions on the copper water pipes, use a tubing cutter to cut both pipes (see Chapter 4: Working with Water Pipe, for instructions for cutting and soldering copper water pipes). You should cut off the copper water pipes about 6 inches (15 cm) above the connectors on the water heater. Keep in mind that you may have to cut the water pipes again to install the new flexible copper connectors.

There should be one screw on the vent draft hood atop the heater. Remove the screw and pull aside the vent pipe.

Now that the old heater is free of the water and gas supply pipes, move it out of the work area. Position the new water heater so the heater gas control valve and gas supply pipe are aligned, and the heater drain valve is to the front of the heater.

Installing a New Strainer

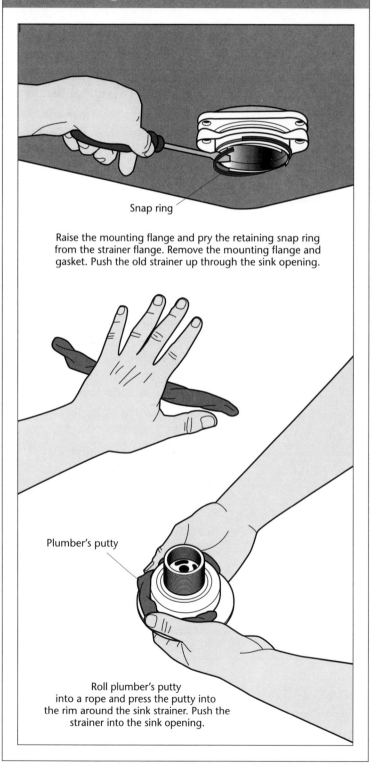

Snap ring

Raise the mounting flange and pry the retaining snap ring from the strainer flange. Remove the mounting flange and gasket. Push the old strainer up through the sink opening.

Plumber's putty

Roll plumber's putty into a rope and press the putty into the rim around the sink strainer. Push the strainer into the sink opening.

Check the hot and cold water pipes to be sure they are aligned with the proper hot and cold sides of the water heater. The water inlets are marked "hot" and "cold" atop the heater.

Check the gas supply pipe to be sure you can connect the pipe to the gas control valve. In many homes there is a steel gas pipe to the heater location, and copper tubing from the steel pipe to the steel fitting at the gas control valve. (If necessary, gently bend the copper tubing so the gas supply lines up with the control valve.)

The flexible connector kit comes with a pair of nipples to connect the water connectors to the heater. If your heater already has male threaded nipples installed, discard the extra nipples. If the hot and cold water inlets have female threads, wrap one end of the nipples with Teflon plumber's tape and use a pipe wrench to turn them water-tight into the water inlet nipples.

Now measure the distance from the nipples to the cut ends of the hot and cold water pipes. Carefully measure and re-cut the pipes to the right length to permit installation of the flexible connectors. Keep in mind when cutting the pipes to length that you must leave the cold water valve in place, or install a new one after

Installing a New Disposal Unit

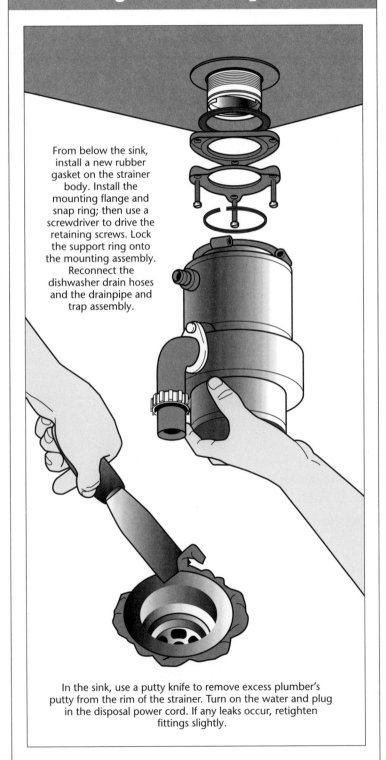

From below the sink, install a new rubber gasket on the strainer body. Install the mounting flange and snap ring; then use a screwdriver to drive the retaining screws. Lock the support ring onto the mounting assembly. Reconnect the dishwasher drain hoses and the drainpipe and trap assembly.

In the sink, use a putty knife to remove excess plumber's putty from the rim of the strainer. Turn on the water and plug in the disposal power cord. If any leaks occur, retighten fittings slightly.

cutting the pipes.

Depending on a number of factors, including existing pipe routes, the height of your new water heater vs. the height of the old one, you may have to do some soldering to hook the pipes to your new heater. See Chapter 4: Working with Water Pipe for soldering instructions. Note that you should not use heat near the hot or cold water connections. Heat will damage the dip tube in the heater. (This tube carries cold water down to the bottom of the heater tank, above the burner.) If you must solder, solder the fitting or adapter to the pipe before you fit the adapter to the water connections of the heater.

Once you have finished hooking up the water pipes, you still have an important piping job to do. There is a temperature and pressure relief valve atop the heater. The T&P valve will open to prevent temperatures and pressures from reaching dangerous levels if the heater malfunctions. To prevent injury to any bystander, you must install a pipe from the temperature and pressure relief valve atop the heater to a nearby floor drain. (See "Water Heater Installation" Illustration on page 80.) Because you should not apply heat to or near the hot and cold water inlets, consider using chlorinated polyvinyl chloride (CPVC) piping at the T&P valve. This plastic pipe can be solvent-welded so there is no danger of heat damage to the heater or adapter. Ask a salesperson at your hardware store to help you get the proper pipe, fittings and solvent.

Apply pipe joint compound to the fitting and reconnect the gas line to the gas control valve. Open the valve on the cold water pipe and refill the water heater. Then turn on the hot water faucet by the laundry tub to let the air escape from the water tank. When the water starts to run without sputtering, the air has escaped, so you can turn off the hot water faucet. When the tank is full and the water has stopped running, check all water pipe connections for leaks. If there are any, tighten the fittings until the leaks stop. If they do not stop, you may have to disconnect the pipes, re-wrap the male pipe threads, assemble and test again.

With pipe joints free of leaks, attach the draft hood to the water heater. Turn on the gas valve. Smell the area around the gas connection to see if you can detect any gas odor. As an extra precaution, coat the gas pipe joints with some soapy water. If bubbles rise in the suds, you have a gas leak. If you do detect a leak, immediately turn off the gas valve, vacate the house and call a service person. If there are no bubbles and you can't smell any gas, you can proceed to start the heater.

Read the heater installation instructions to see how to turn on the heater. Some heaters have a standing pilot light that must be lit with a match. Others have a piezo-electric ignition. See the manufacturer's pilot lighting instructions for your heater. For the standing pilot light, instructions are to remove both inner and outer doors from the heater. Turn the temperature setting counterclockwise to its lowest setting. (See "Thermostat" Illustration on page 81.) Turn the gas cock knob clockwise to the "off" position. Wait five minutes to clear out any gas, again checking for gas odor.

Turn the gas cock knob to the "pilot" position. Push the red button down, and hold a match to light the pilot. After the pilot ignites, hold the red button down for about a minute. Release the red button and the pilot should stay lit. If it does, replace the inner and outer doors. Turn the gas cock knob counterclockwise to the "on" position and turn the heater temperature dial to the desired setting. If the pilot light refuses to stay lit, turn off the gas valve and call a service person.

Note that many people have been badly scalded, some fatally, by high water temperature settings. If there are young children or elderly persons in your household, limit the water heater's temperature setting to 120 degrees Fahrenheit (49°C). If you live in an adult household, start with a low 120-degree setting until you test the water, then raise the thermostat a notch at a time until you reach the desired setting. In no case, should the temperature be set above 130 degrees Fahrenheit (54°C). At 130 degrees Fahrenheit, hot water can produce serious burns in as little as 30 seconds of exposure. (See "Temperature and Burn Times" Box on page 81.)

REPLACING A GARBAGE DISPOSAL

Replacing a garbage disposal is within the abilities of the home handyperson. Note that the procedure for replacing a disposal unit is similar to repairing leaks around the sink strainer, the water hoses connecting the dishwasher to the disposal, and/or a leaking trap assembly. You simply perform the same steps except that you reinstall the old disposal unit, rather than replacing it. While you have the disposal/drain assembly apart, do some preventive maintenance. Replace the dishwasher water hoses; clean and renew the plumber's putty around the strainer, and install new slip nut washers. Use silicone sealant at drain and trap slip nuts to ensure leak-free operation.

To replace the disposal unit, first turn off the water supply valves to the dishwasher, and unplug the power cord to the disposal. Keep a cake pan or plastic pail handy and set it under the trap, disposal or dishwasher water hoses to catch any water.

Now use channel-type pliers to loosen the slip nuts on the trap assembly. Set the trap assembly aside and examine it to be sure it is free of debris. Remove the garbage disposal drainpipe. (See "Placing a Garbage Disposal" Illustration on page 84.) Use a screwdriver to loosen the screws on the dishwasher hose clamps, and pull the hoses off their connections at the garbage disposal.

Place the pail under the disposal unit to catch any water. Then put one hand under the disposal unit and with the other hand turn the lower support ring to unlock it from the mounting assembly. Remove the screws from the mounting assembly and push the mounting flange upward. Use a flat-blade screwdriver to pop the snap ring off the strainer flange. (See "Removing the Old Disposal Unit" Illustration on page 85.) Push the strainer back up through the sink opening.

Roll a bead of plumber's putty into a rope, and place the putty around the rim of the new sink strainer. Push the strainer back down through the sink opening. (See "Installing the Strainer" Illustration on page 86.) From below the sink, slip the rubber gasket onto the strainer and install the disposal mounting flange and snap ring. Replace the retaining screws.

Place one hand under the new disposal and turn the lower support ring to lock it onto the mounting flange. (See "Installing the New Disposal Unit" Illustration on page 87.) Use a flat-blade screwdriver to reconnect the dishwasher water hoses. Install the disposal drainpipe and connect the trap assembly. Turn on the water and plug in the disposal power cord. Test all fittings for leaks: retighten where necessary.

CHAPTER 9

Remodeling the Bath

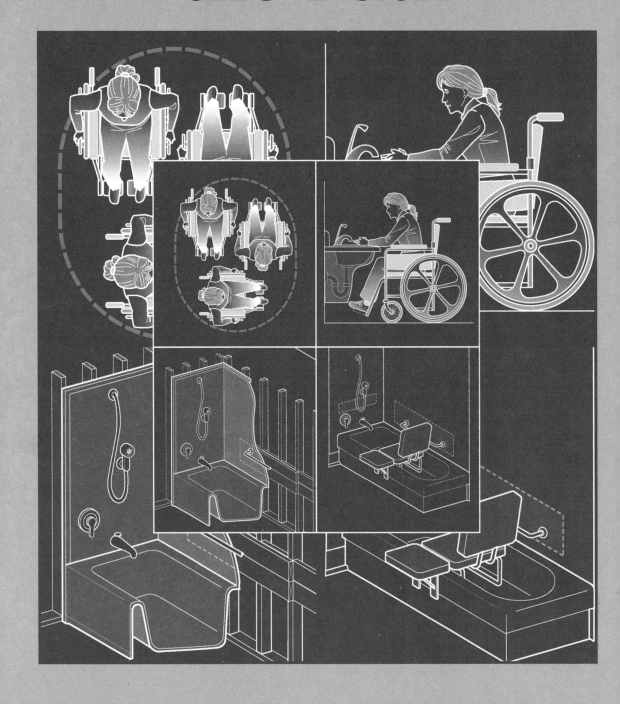

The bathroom is the most remodeled room in the house—and potentially the most dangerous. Wet surfaces such as floors and tiled walls can cause many falls and injuries. Bathrooms may be small and cramped, making entrance and access difficult for the elderly or disabled. If you decide to remodel the bath, consider adding features that will make the bathroom more safe and convenient for all users, and barrier-free for those who have physical limitations. Having a barrier-free bathroom in the house not only makes the room more convenient for the present occupants, it can be a real plus when it comes time to sell. Estimates for the state of Minnesota, for example, are that one out of four families has a member who has one or more physical limitation. This proportion will only increase as the baby boomer generation ages.

Begin planning by considering access to the room. In many older homes, bathroom doors measure 2 feet, 6 inches (30 in. or 0.76 m), which makes them too narrow for wheelchair entry. If you have a narrow door like this, think about replacing it with a 2-feet 8-inch (32 in. or 0.835 m) door or even better with a 3-feet-wide door (36 in. or 0.91 m). Room is required to swing a door, and most bathroom doors swing inward. Consider installing a door that swings out, or even a sliding pocket door, to eliminate the space used by an inward-swinging door. (See "Bathroom Plan for Barrier-Free Entry" Illustration on page 92.)

An adult-sized wheelchair is 26 inches (66 cm) wide, and an additional allowance of approximately 6 inches (15.24 cm) must be made to accommodate the user's elbows. To make a full U-turn in a wheelchair requires an open floor space of 5 feet (1.52 m) by 6 ½ feet (1.98 m). Allow as much space as possible to facilitate wheelchair use. Keep in mind that even those recovering from injuries or surgery may find it difficult to enter the bathroom and move around in it. Wall-hung sinks and toilets can be used to open up access to the room. (See "Wheelchair U-turn" Illustration below.)

Next, consider the plumbing fixtures. Choose faucets that have a single lever, rather than a knob. The single lever not only offers both water volume and temperature control in one device, but it is also easier to grip and can even be operated by a push of the hand or arm. This is a real plus for those with limited mobility. When you replace any faucet in the house,

Wheelchair U-turn

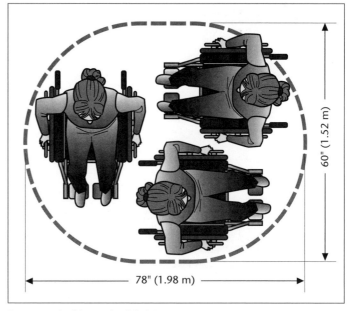

60" (1.52 m)

78" (1.98 m)

Space required for a wheelchair U-turn.

Bathroom Plan for Barrier-Free Entry

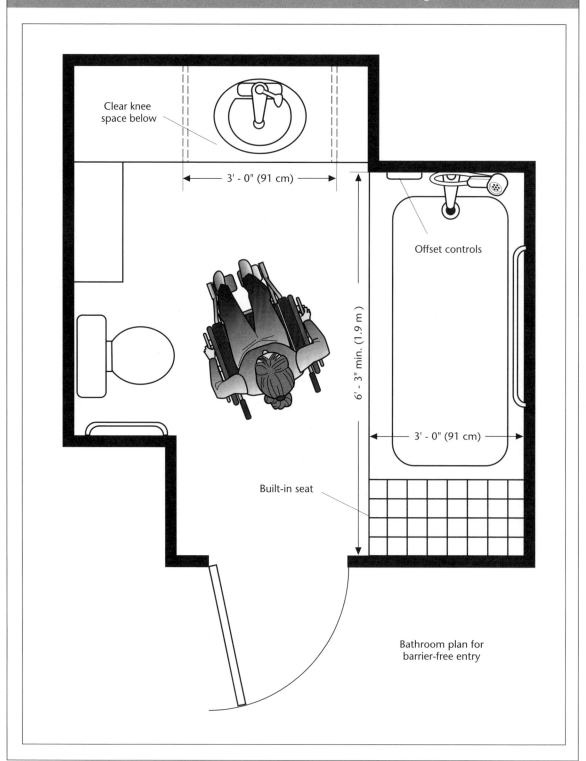

Clear knee space below

3' - 0" (91 cm)

Offset controls

6' - 3" min. (1.9 m)

3' - 0" (91 cm)

Built-in seat

Bathroom plan for barrier-free entry

Standard Bathtub With Safety Features

Hand-held shower

Wall reinforcing for grab bars

Offset water valve

Removable tub seat

Standard Bathtub With Built-In Seat

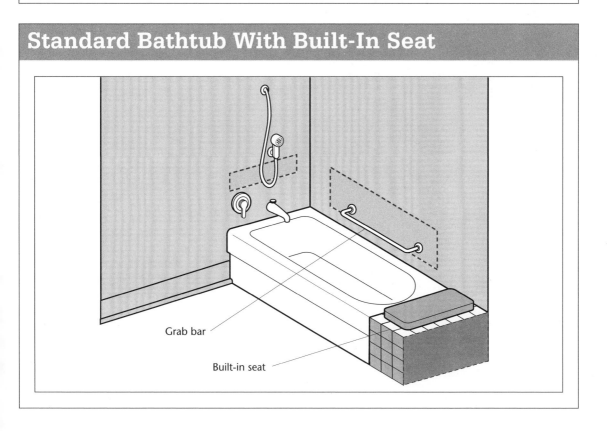

Grab bar

Built-in seat

Wall Reinforcement for Bathtub Grab Bars

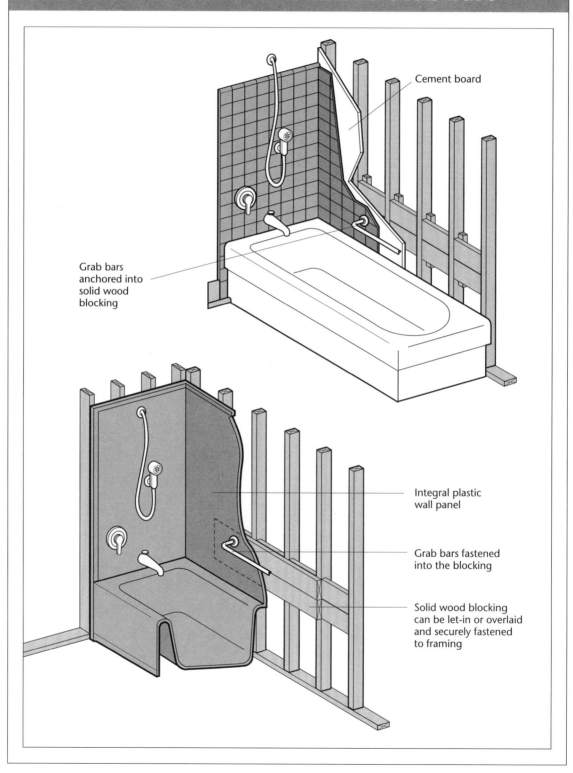

Cement board

Grab bars
anchored into
solid wood
blocking

Integral plastic
wall panel

Grab bars fastened
into the blocking

Solid wood blocking
can be let-in or overlaid
and securely fastened
to framing

Barrier-Free Sink

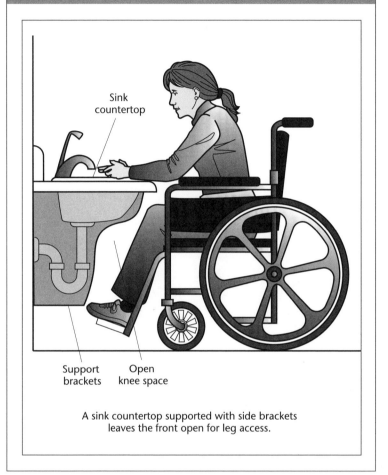

Sink countertop

Support brackets

Open knee space

A sink countertop supported with side brackets leaves the front open for leg access.

built with a vinyl seal around the door, and have only a low vinyl threshold at the floor. This vinyl threshold permits a disabled person to roll into the shower while seated in his or her wheelchair.

For tub access, build a seat at the end of the tub. This permits those with limited mobility to sit down at the end of the tub, swing their legs into the tub, and hold onto grab bars while easing themselves into the bath. (See "Bathtub with Built-in Seat" Illustration on page 93.)

Pedestal sinks allow legroom for wheelchair access. An alternative is to remove the front of the vanity cabinet so a wheelchair user can roll the front of the chair under the countertop. (See "Barrier-Free Sink" Illustration at left.)

A wall-mounted toilet or vanity sink can be installed at any height. If necessary, the toilet can be installed to match the height of the wheelchair seat, so movement is easier between the two seats. To aid those who are infirm, grab bars can also be installed by the toilet.

choose one with a single lever.

Many injuries are sustained from falls in a slippery tub or shower stall. All tub and shower areas should be equipped with grab bars. Grab bars are available in many configurations and can be added to an existing bath. If you are remodeling, plan where you will locate grab bars and nail 2 × 6 blocking between studs so you will have a solid support to screw the grab bars to. To be sure the bar will support the weight of a falling person, you must use a stud locator to find the studs and use long screws to anchor the bar to solid framing. (See "Wall Reinforcement for Bathtub with Grab Bars" Illustration on page 94.)

Preformed fiberglass shower stalls and bathtubs are available with flexible extension hand-held showers, either fixed or folding seating, and built-in grab bars. Some shower stalls are

PART IV

ADVANCED TECHNIQUES

CHAPTER 10

Adding a Half Bath

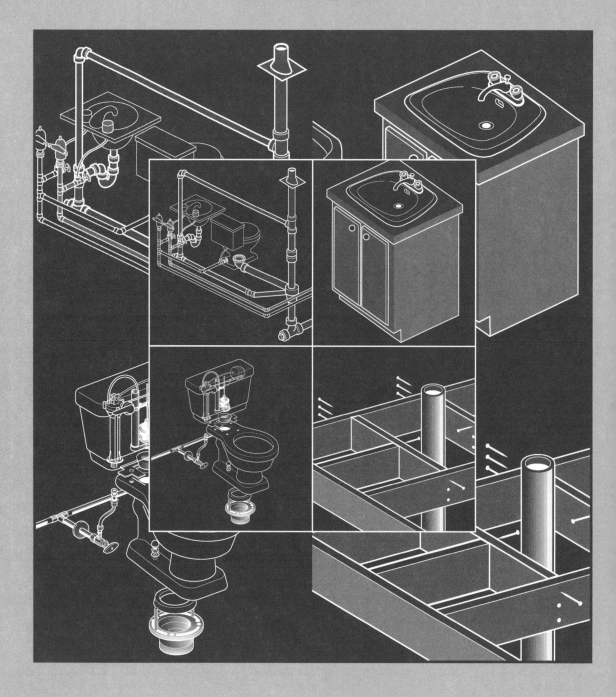

The bathroom can be a busy room, particularly in the morning when the family is trying to get off to work or school. To reduce morning traffic, consider adding a half bath to your home. A half bath includes a toilet, a sink and a medicine cabinet with mirror. A half bath will not only make your life more convenient but will also add value to your home at resale time. Adding a half bath is an addition that will be sure to return your investment.

Although a larger space is desirable, a half bath can be squeezed into a space as small as 3 × 4 feet (0.91 × 1.22 m). This means you may be able to fit the bath into the end of a long hallway, a large closet or the corner of a large bedroom. To simplify plumbing work, locate the new half bath as close as possible to existing plumbing. (See "Half Bath Plumbing Layout" Illustration on page 100.)

To fit the half bath in a small space plan on using such space-savers as a 12 × 12 inches (30.48 × 30.49 cm) sink, a lowboy or low silhouette toilet, and a sliding or pocket door. Because the pocket door slides into the wall cavity, it does not require space to swing like a conventional hinged door. If the space available is along an outside wall you can add a window

that will provide light and ventilation while making the room seem larger. If the half bath will be built within interior walls only, consider adding a skylight or roof window, or provide ventilation using a power vent system. The power vent can vent via the ceiling, exiting through the outside wall or the roof.

With all the above points in mind, draw a sketch of the half bath. The toilet will measure about 20 inches (50.8 cm) wide and will extend about 28 inches (71.2 cm) from the wall. A wall-hung sink may extend 18 inches (45.8 cm) from the wall. Consider buying the toilet and sink before finalizing your plan, so you can be sure the fixtures will fit into the space allotted. Sink options include a sink hung by brackets on the wall, a pedestal sink, or if enough space is available, a sink set into a vanity cabinet. (See "Sink Options" Illustration on page 102.) Because it is the larger fixture, plan the location of the toilet first, then fit the sink in where possible.

When your plan is complete, begin the demolition work necessary to install new plumbing, vents and electrical service to the new room. Demolition of existing walls means removing base trim, plaster or wallboard in those areas that must be disturbed

to complete the job. For example, to install the vent stack you may have to remove the trim and plaster or wallboard between two wall studs. To install the new toilet drainpipe you may have to cut out a portion of a floor joist. If this is necessary you must cut a header the same size as the floor joist (2 × 8 or 2 × 10) long enough to span the distance between the two adjoining joists. Nail this header to the two adjoining joists and through the end of the cut floor joist. (See "Cutting a Floor Joist" Illustration on page 100.)

Plastic pipe should be the pipe of choice for new plumbing. Plastic pipe is lightweight, flexible (so it can be routed through holes in framing members) and can be joined with solvent to avoid soldering with an open flame. Install the new plumbing as shown in the "Half Bath Plumbing Layout" Illustration on page 100. You must cut the existing plumbing pipes, install T-fittings into cold and hot water supply, and run new pipe to the toilet tank and the sink. The hot water pipe at the fixture is always the pipe on the left-hand side; a riser from the cold water pipe also serves the toilet. It is easier to solvent weld several pipes and fittings together before positioning them. Solvent weld an

Half Bath Plumbing Layout

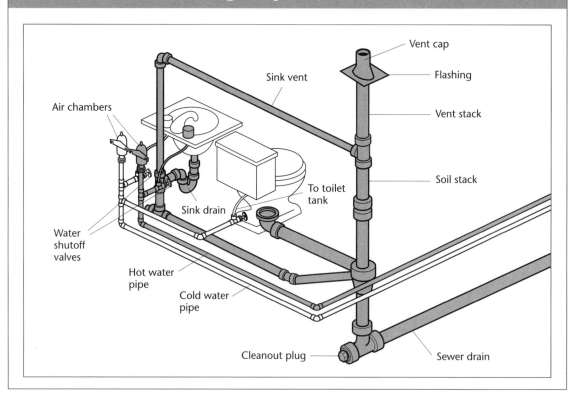

Labels: Vent cap, Flashing, Vent stack, Sink vent, Soil stack, Air chambers, To toilet tank, Sink drain, Water shutoff valves, Hot water pipe, Cold water pipe, Cleanout plug, Sewer drain

Cutting a Floor Joist

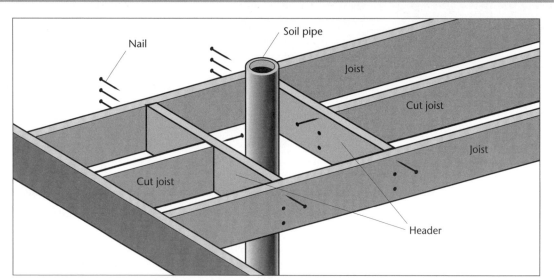

Labels: Nail, Soil pipe, Joist, Cut joist, Joist, Cut joist, Header

If you must cut a joist to install the toilet drain, nail in a header to transfer the floor's weight to the two adjoining floor joists.

Toilet Components

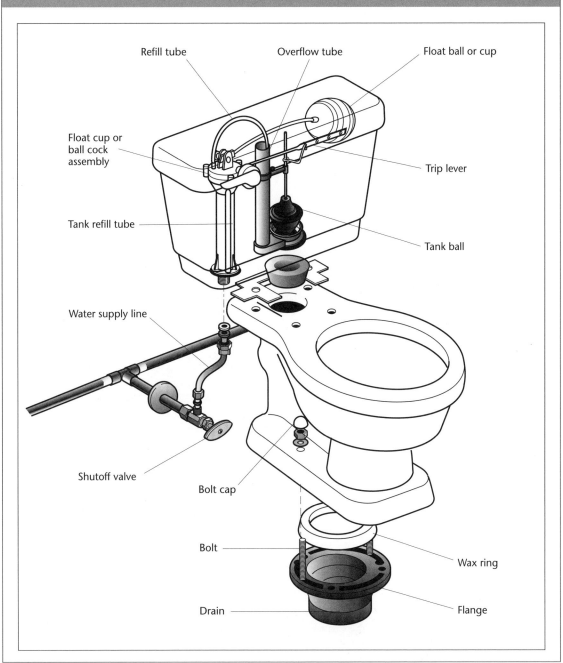

Refill tube

Overflow tube

Float ball or cup

Float cup or ball cock assembly

Trip lever

Tank refill tube

Tank ball

Water supply line

Shutoff valve

Bolt cap

Bolt

Wax ring

Drain

Flange

assembly of pipe and lift it into place. Use plastic pipe hangers to hold the pipe assembly in place while you add further fittings.

Use 3-inch (7.6 cm) plastic pipe for the vent stack. The vent stack should be installed in a wall cavity between two studs. Cut 3

inch (7.6 cm) holes through the top and bottom 2 × 4 plates. Cut a hole in the roof and install the vent pipe through it. Flash the vent

Sink Options

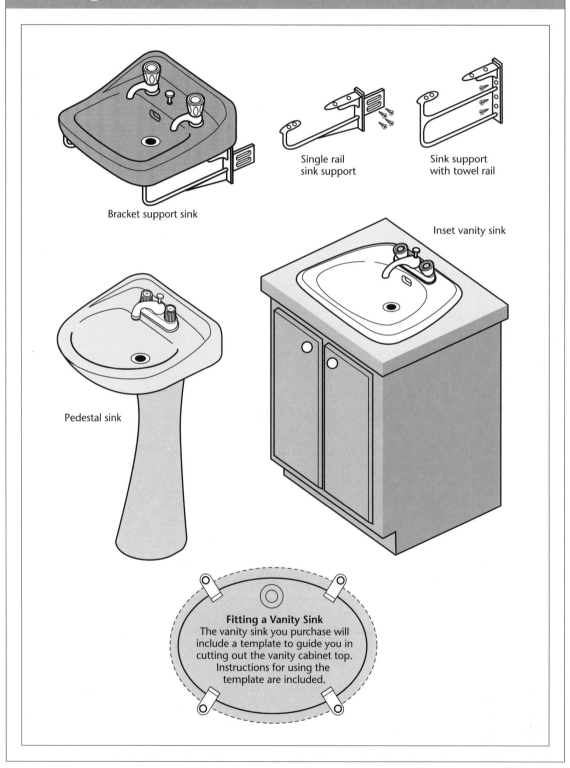

Bracket support sink

Single rail
sink support

Sink support
with towel rail

Inset vanity sink

Pedestal sink

Fitting a Vanity Sink
The vanity sink you purchase will
include a template to guide you in
cutting out the vanity cabinet top.
Instructions for using the
template are included.

Cutting Cast-Iron Drainpipe

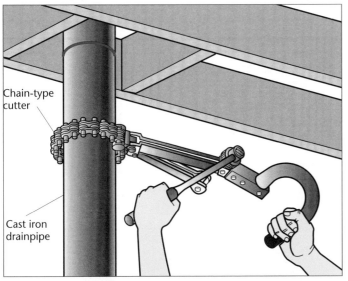

Chain-type cutter

Cast iron drainpipe

Rent or borrow a chain-type pipe cutter and cut out a section of the existing soil stack. When the section is cut free remove it and install a sanitary T to receive the new toilet drainpipe.

Using No-Hub Connectors

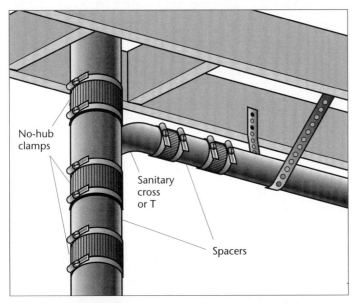

No-hub clamps

Sanitary cross or T

Spacers

Center no-hub connectors over the joints between pipes or fittings. Use a screwdriver or ratchet wrench to tighten the clamp bolts.

pipe and install a cap to prevent water from entering through the top of the vent hole.

Use ½-inch (1.27 cm) plastic T-fittings and pipes to join the new water supply lines to the existing lines. To cut into the cast-iron soil stack and connect the toilet drainpipe to the existing stack, rent a chain-type pipe cutter. This tool has a chain that incorporates cutting teeth in the chain. Wrap the chain around the soil stack, and secure the end of the chain to the handle. Then turn the tool so it revolves around the stack. Continue to turn and tighten the pipe cutter until the pipe is cut free. (See "Cutting Cast-Iron Drainpipe" Illustration at left.) Remove the cutout section of soil stack and use no-hub connectors to install a sanitary T. No-hub connectors consist of a neoprene sleeve or tube with stainless steel clamps attached. Use a socket wrench to tighten the clamps on the joints. (See "Using No-Hub Connectors" Illustration at left.) Remember when cutting and installing the new toilet drainpipe that it should slope downward from the toilet to its connection into the soil stack at a rate of ¼ inch (6.35 mm) per running foot (30.48 cm).

CHAPTER 11

A Wall-Hung Sink

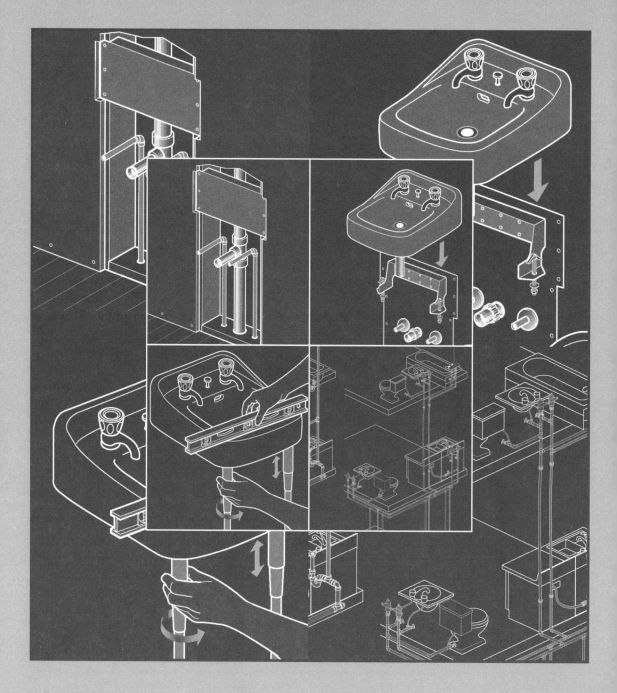

There are still many of wall-hung sinks around, relics of the past before vanity cabinets became popular. If you want to replace an existing wall-hung sink, or install a wall-hung model to save space in a half-bath, this chapter will show you how.

If you are installing a sink at a new location, the first step will be to remove the old wall finish as shown in the illustration. This will open up the stud cavity for installing the pipes. Next, run water lines and a drain-pipe to the new sink location. Cut the hot and cold water supply pipes near the new sink location and insert T-fittings into them. Now run new pipes from the T-fittings to the sink location. Remember the hot water pipe will be on the left, the cold water pipe on the right.

Cut a length of 2 × 10 long enough to span the distance between the two studs, and nail the 2 × 10 blocking to the studs. The top of the 2 × 10 blocking should be set about 35 inches (88.9 cm) above the floor. This 2 × 10 will supply support for the new sink. (See "Installing the Bracket" Illustration at right.) With piping and blocking installed, cut new wall finish material such as wallboard to fit the opening, marking and drilling holes for the drain

Installing the Bracket

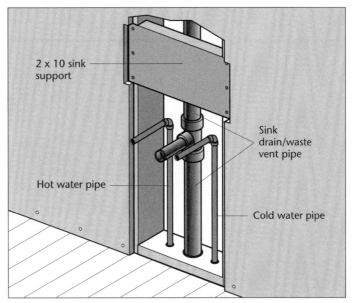

2 x 10 sink support

Sink drain/waste vent pipe

Hot water pipe

Cold water pipe

To install a new wall sink, remove the finish wall material between wall studs at the chosen location. Install a section of 2 × 10 between the studs to provide support for the sink's hanger bracket.

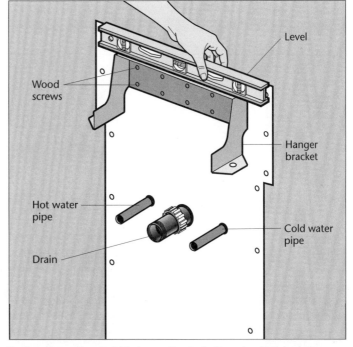

Level

Wood screws

Hanger bracket

Hot water pipe

Cold water pipe

Drain

Replace the wall material between the studs. Use a level to check the hanger bracket and attach the bracket to the 2 × 10 using wood screws.

Mounting the Sink

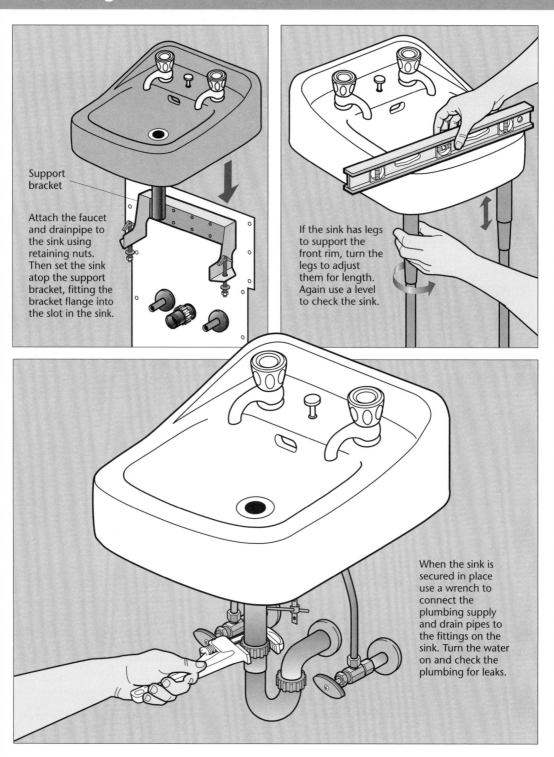

Support bracket

Attach the faucet and drainpipe to the sink using retaining nuts. Then set the sink atop the support bracket, fitting the bracket flange into the slot in the sink.

If the sink has legs to support the front rim, turn the legs to adjust them for length. Again use a level to check the sink.

When the sink is secured in place use a wrench to connect the plumbing supply and drain pipes to the fittings on the sink. Turn the water on and check the plumbing for leaks.

DWV System and Water Supply System

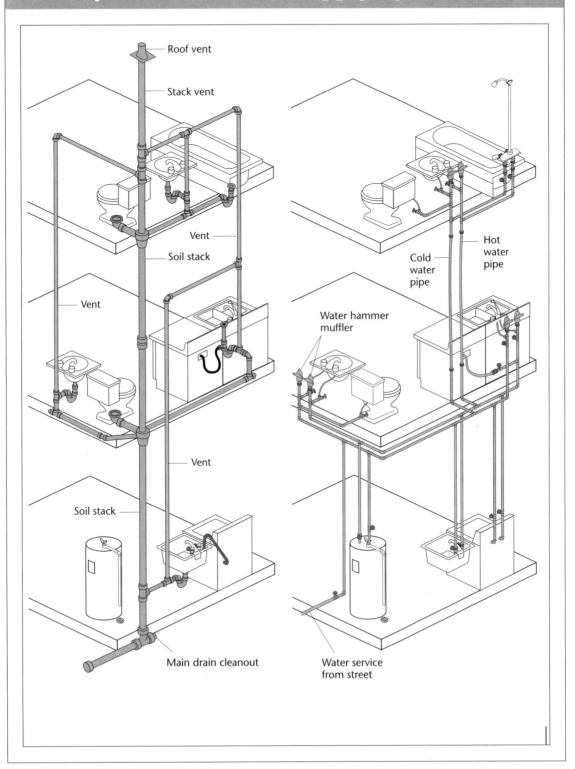

Roof vent

Stack vent

Vent

Soil stack

Vent

Vent

Soil stack

Main drain cleanout

Water hammer muffler

Cold water pipe

Hot water pipe

Water service from street

Pedestal Sink

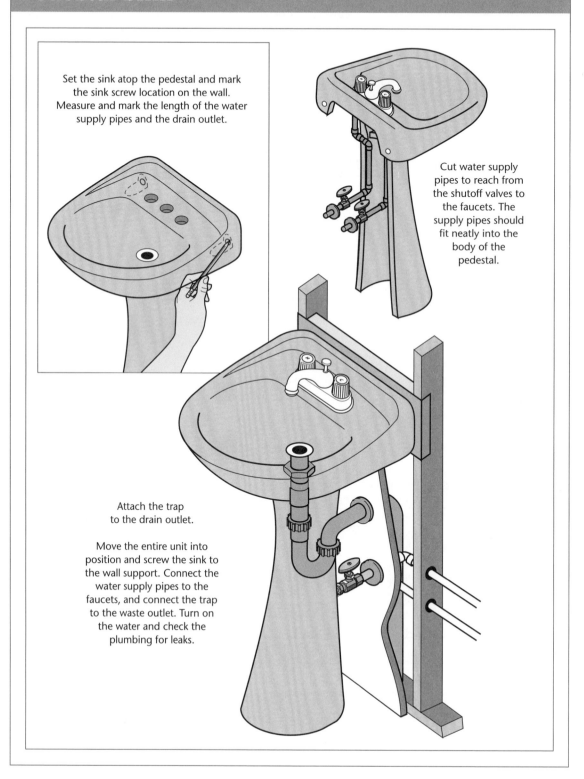

Set the sink atop the pedestal and mark the sink screw location on the wall. Measure and mark the length of the water supply pipes and the drain outlet.

Cut water supply pipes to reach from the shutoff valves to the faucets. The supply pipes should fit neatly into the body of the pedestal.

Attach the trap to the drain outlet.

Move the entire unit into position and screw the sink to the wall support. Connect the water supply pipes to the faucets, and connect the trap to the waste outlet. Turn on the water and check the plumbing for leaks.

and pipes. Nail the finish material in place and tape and finish the cracks and nail holes.

Check the instructions included with the sink to find the recommended hanger bracket height. Use wood screws to screw the hanger bracket to the 2 × 10 backing. Check as you screw the hanger bracket on that it is level. (See "Installing the Bracket" Illustration on page 105.)

Turn the sink over and use the retaining nuts to secure the faucet and the drain pipe body in position on the sink. Turn the sink upright and slip the sink over the hanger bracket. The flange of the bracket slips into a slot in the sink. (See "Mounting the Sink" Illustration on page 106.) Some models have toggle bolts to help secure the flange in the slot. Note that wall-hung sinks are serviceable but cannot support a great deal of downward pressure.

Pedestal Sink

As mentioned in Chapter 10, Adding a Half Bath, a pedestal sink is yet another space-saving option. The pedestal offers plenty of support, does not require much space, and is considered an attractive addition to a bathroom.

The height of the risers, or water supply pipes, is dictated by the height of the pedestal sink you choose. To begin installation, set the sink atop the pedestal. Measure the length of the hot and cold water risers you will need as well as the length of the drain outlet. Now set the unit so the sink is against the wall, and mark the screw holes where the screws will go into the wall support. Plan the plumbing so the risers will fit into and be concealed by the pedestal. (See "Pedestal Sink" Illustration on page 108.)

CHAPTER 12

Wells and Septic Systems

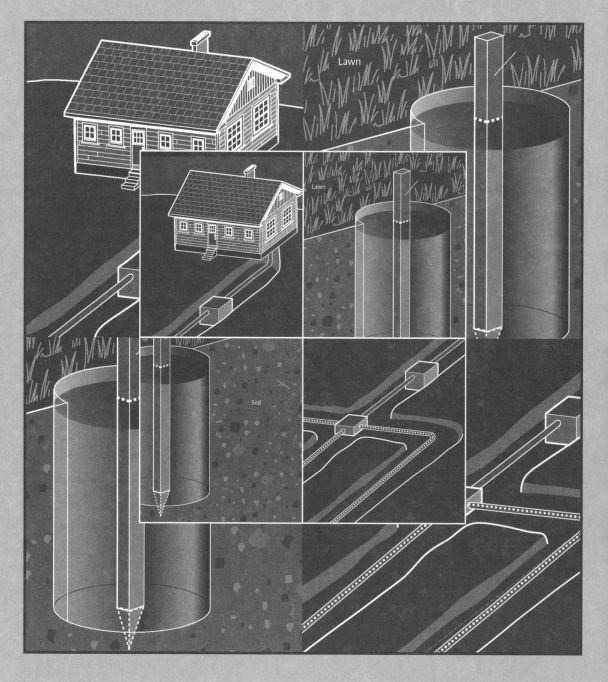

If you live beyond city utility lines your water supply must come from a well, and you must build your own sewage disposal system. Repairs to both these systems must usually be done by a professional, but there are steps you can take to reduce service calls and clues that may warn you that your systems need to be serviced.

WELLS

There are a few well repairs within the abilities of the average handyperson. Listen to see how your well pump is operating. The water tank should contain air. When the pump is on, the air is compressed in the top of the tank until the water pressure reaches the proper level. When the pressure falls, the pump cycles on and runs until the operating pressure is reached. If you hear the well pump cycle on and off frequently, the water tank may be water locked. This means there is no air in the tank to pressurize the system. The solution is to drain the water from the tank so it can fill with air. Shut off the pump switch and the water supply from the well pipe and drain the tank. Now shut off the tank drain valve, open the supply valve and turn on the pump switch. The water should now come up to normal pressure as

indicated on the pump pressure gauge. If the water tank is not holding air, you will have to repeat this operation when the pump again cycles on intermittently. Some tanks contain an air bladder that prevents this condition. Consider replacing your old water tank with a new one that incorporates an air bladder.

Have your well water checked periodically to be sure it is safe to drink. Wells sometimes are contaminated by surface absorption of agricultural chemicals or other chemicals, so have a laboratory do a check once a year.

Also check the water for sediment. If you find sand in the water, it's likely that the well screen at the bottom of the well pipe has failed to work and sand is entering your system from the aquifer below. Have the well checked by a service person. If sand is present, the service person will have to pull the well pipe and install a new screen.

If fixtures such as sinks and bathtubs develop rust or other stains, your water may have a high iron or mineral content. If this is the case, install a water filter in the line to filter these materials from the water.

SEPTIC SYSTEM

A septic system is composed of the main drainpipe from

the house and a septic tank. From the septic tank, a drainpipe extends to a distribution box. From the distribution box, perforated pipes are laid to form a leach or seepage field. (See "Example of a Septic System" Illustration on page 112.)

The septic tank may be made from preformed concrete, fiberglass or steel. Septic tanks are available in sizes ranging from 500 to 1500 gallons (1,892 to 5,678 liters). The size of tank needed is calculated based on the number of persons in the house and the estimated water usage. If you are building a new house, it is advisable to opt for a large septic tank.

Sewage waste flows from the house into the septic tank. There the solid wastes settle to the bottom of the tank to form sludge. The liquids rise to the top of the septic tank and flow out of the tank through the pipe to the distribution box. Grease and household detergents or other chemicals will float as a scum at the top of the wastewater.

At the distribution box, the wastewater is directed through two or more perforated pipes where the water is absorbed into the soil and eventually evaporates.

The need for a distribution box and seepage field pipes depends on the soil. The seepage field trenches

Example of a Septic System

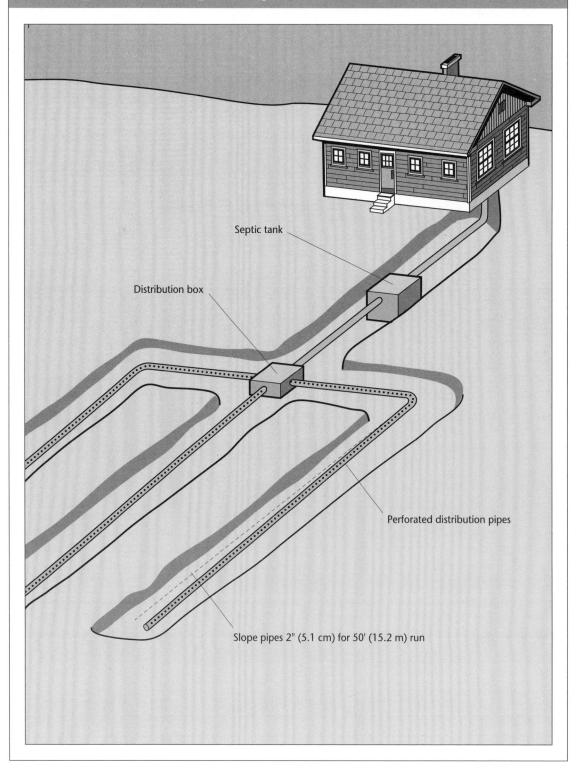

Septic tank

Distribution box

Perforated distribution pipes

Slope pipes 2" (5.1 cm) for 50' (15.2 m) run

Soil Percolation Test

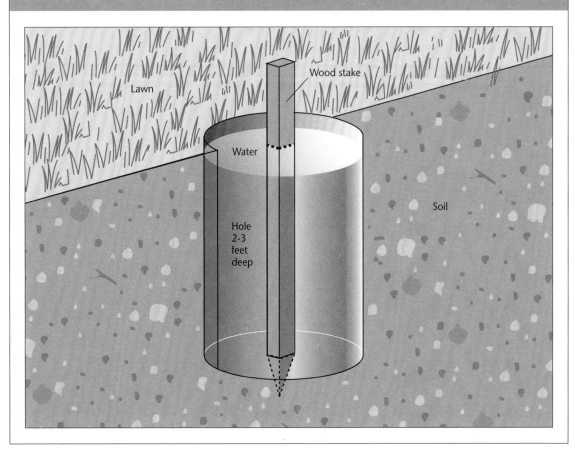

Lawn

Wood stake

Water

Hole 2-3 feet deep

Soil

are often filled with gravel for better water absorption. If the soil is sandy or sandy loam, it will absorb the wastewater well and the size of the leach field can be reduced. If the soil is heavy clay, water absorption will be poor and more distribution pipes and gravel fill will be needed. The septic tank installer will run a percolation test on your soil to test its ability to quickly absorb water. (See "Soil Percolation Test" Illustration above.)

The percolation test is done by digging a hole in the ground 2 to 3 feet (0.6 to 1 m) deep. Fill the hole with water and let it stand until the water is absorbed. Refill the hole and again wait until the water is absorbed. Now place a stake in the bottom of the hole, extending to above ground level. Fill the hole with water and mark the water level on the stake. Wait one hour and measure from the water-level mark on the stake down to the new water level. If the water level has dropped 1 inch (2.5 cm) or more in one hour, you have good soil percolation. If it has dropped less than 1 inch (2.5 cm) you will need more seepage pipes and gravel for absorption.

SEPTIC TANK MAINTENANCE

Because a septic system is limited by size in the amount of sewage it can handle, your aim should be to reduce the amount of wastewater flowing from the house into the system. Install low-flow showerheads and water-conserving low-flush toilets in the house.

Take short showers. Do not leave water running in the sink when washing dishes or vegetables. Instead, fill the sink and rinse the dishes or vegetables in the standing water.

When shopping, consider the effect that laundry detergents, bleach and other household supplies may have on the septic system. Many common chemicals, including bleach, will affect the bacteria count in the septic system. The bacterial action is needed to break down the solids into sludge, so consider whether using any chemical is truly necessary. Note that products are available that are advertised to improve the bacterial action in a septic system; ask your service person if he or she recommends using these products.

As noted in Chapter 5: Drain Maintenance, you should never use the toilet as a garbage can. Toilet tissue is made to degrade when wet, but other paper products such as soap wrappers are not. Keep a wastebasket under the vanity and dispose of these other bathroom products in the wastebasket, not in the toilet.

Have a service person pump the septic tank periodically to remove the sludge. He or she will advise you of how often this should be done. If sludge is allowed to accumulate until the septic tank is filled, sewer solids will flow over the sludge and directly into the distribution pipes. The solids will plug these pipes, which are intended to disperse wastewater only. Once plugged, the distribution pipes will have to be removed and replaced, an expensive project. You'll not only need a new septic system, but you will need a new lawn, because it will have to be dug up so the distribution pipes can be reached.

CHAPTER 13

Installing a Sprinkler System

Much of the work associated with owning a home revolves around yard maintenance. If you want to have a nice lawn and garden, but wish to limit the necessary upkeep and hard work, install a lawn sprinkler system. This system will not only save labor, but by managing the system you can conserve water and reduce water bills. For example, the most common grass in the northern half of the U.S. is bluegrass. Bluegrass requires a total of two inches (5.1 cm) of water every two weeks. This amount includes both lawn watering and rainfall. By monitoring the weekly rainfall and supplementing the rain with watering you can meet the lawn's water requirement of two inches without wasting water.

Installing a lawn sprinkler system is a fairly simple plumbing project. The first step is to check the water pressure to be sure you have enough pressure to operate the system. Your utility company or a plumber can attach a pressure gauge to the outside hose bib to make this check. When you are sure your water pressure and volume are sufficient to handle the sprinkler system, proceed to planning.

To begin the planning step, first measure the lawn and make a drawing of the lawn area. Mark special areas such as flowerbeds, walks, storage sheds and large trees. Take this drawing along when shopping for the sprinkler system. Using plastic pipe and fittings, the plumbing is straightforward. The configuration and dimensions of the yard will determine the need for pipes, fittings and sprinkler heads.

At the home center, first choose the sprinkler heads. The heads are available to water in quarter-circle, half-circle, full circle and square patterns. The sprinkler pattern you choose will depend on the shape of the area to be watered and the capacity of the heads. The area covered by each sprinkler head will determine the location and number of sprinkler heads needed. (See "Lawn Sprinkler System" Illustration on page 118.)

While planning your sprinkler system you must also decide on which type of plastic pipe or tubing you will use. For some products manufacturers advise that you buy flexible tubing and use clamps to join the tubing and the sprinkler risers. The flexible plastic tubing is available in 100-foot-long rolls (30.48 meters). These rolls are available at home centers at a cost of about $10. The flexible tubing can be curved slightly, but avoid sharp bends or curves. Any sharp bend or curve in the supply pipe will decrease the water pressure to the sprinkler heads.

Other manufacturers advise using rigid plastic pipes in 10-foot lengths (3.05 meters) and connecting the system together using the solvent-welding method.

You must also decide whether you will have an automatic or a manual sprinkler system. The automatic control timer can be set to water at whatever interval you choose. This lets you water the lawn in a "set it and forget it" mode. Your lawn will be divided into zones to meet water demand. The control manifold will open and close the zone pipes alternately, sprinkling a portion or zone of the lawn. A tradeoff for automatic systems is that the sprinkler will turn on at preset intervals, even during a heavy rain, which wastes water.

If you choose to install a manual sprinkler system you must turn on the valve and water one zone at a time. This is not as convenient as the automatic system, but you will still eliminate the task of dragging long garden hoses around and setting sprinklers that will water only a small area at a setting, and then must be moved.

Whichever system you choose, the plumbing is much the same. First, tap

Lawn Sprinkler System

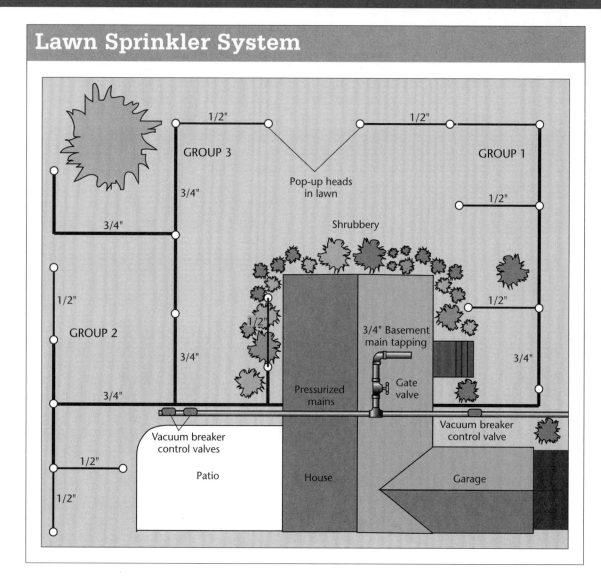

into the cold water line inside the house. To ensure maximum pressure make this connection as close as possible to the water meter. Install a T into the line as instructed in Chapter 4, Working with Water Pipe. You can make the new connection with either plastic or copper fittings. Also, install a gate valve in your new pipe so the water can be shut off from the basement or crawl-space. Drill a hole large enough to install the pipe through the outside wall. When you have installed the pipe through the hole, caulk around the pipe to ensure a weather seal. (See "Connecting to House Plumbing" Illustration on page 119.)

Outside, you must install a control manifold, either an automatic manifold if automatic is your choice, or a manual control manifold. You must install control valves that include a vacuum breaker. The vacuum breaker is an anti-siphon feature that prevents a cross-connection. A cross-connection allows water from the sprinkler system to be drawn back into the house water system potentially contaminating the house water supply. The vacuum breaker valves should be installed so they

Connecting to House Plumbing

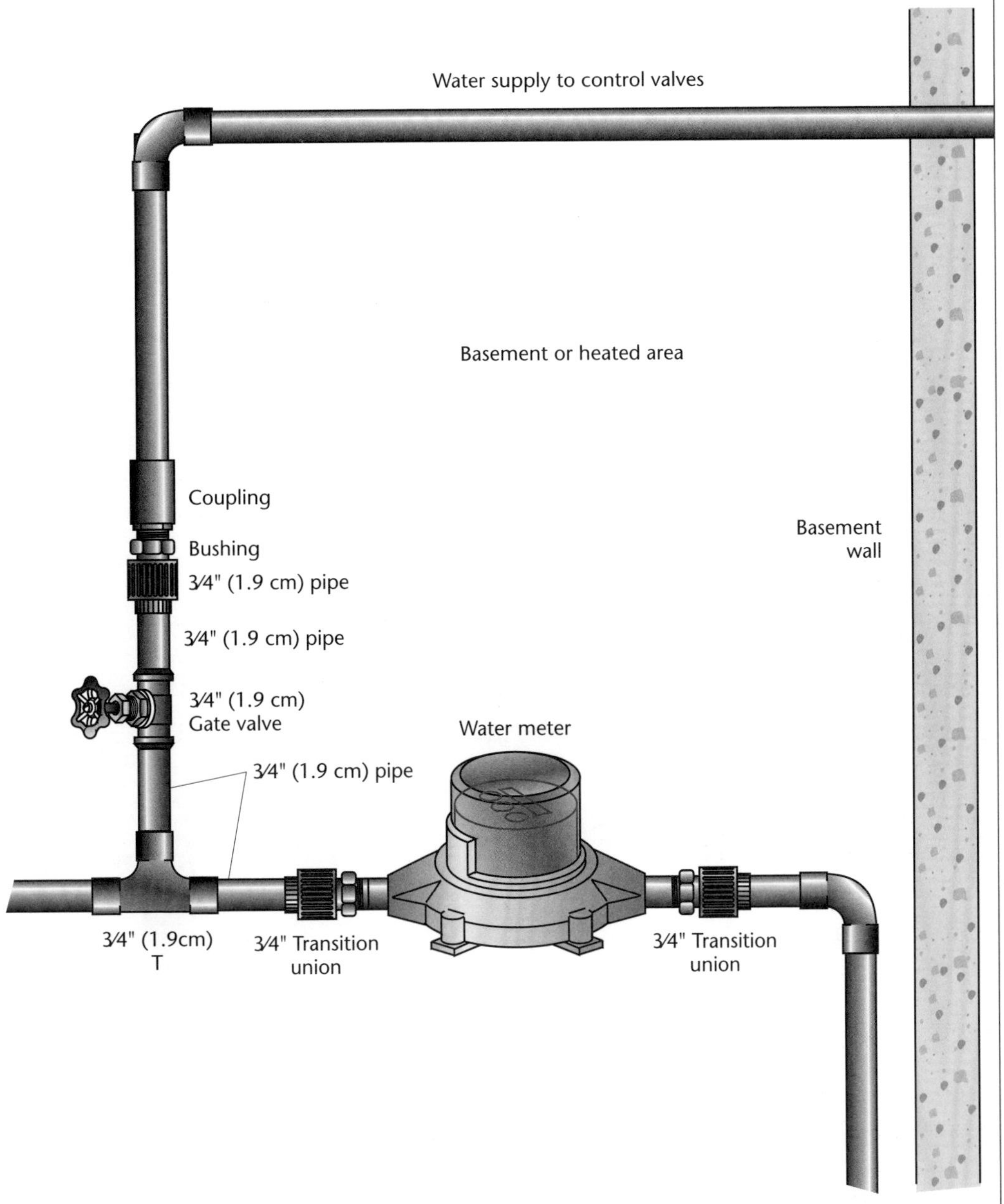

Water supply to control valves

Basement or heated area

Basement wall

Coupling

Bushing

3⁄4" (1.9 cm) pipe

3⁄4" (1.9 cm) pipe

3⁄4" (1.9 cm)
Gate valve

3⁄4" (1.9 cm) pipe

Water meter

3⁄4" (1.9cm)
T

3⁄4" Transition
union

3⁄4" Transition
union

Control Manifold

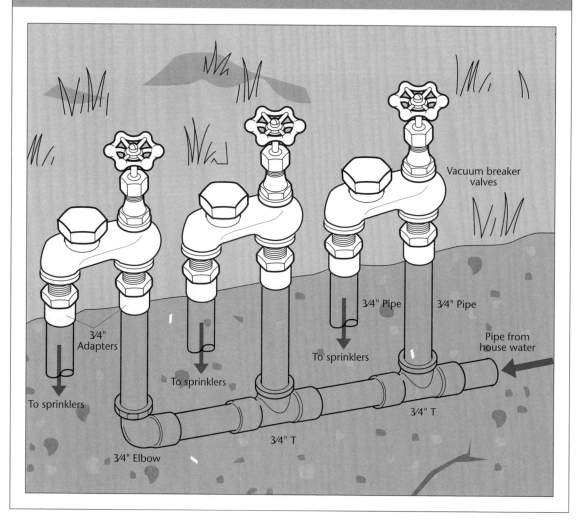

Vacuum breaker valves

3⁄4" Pipe   3⁄4" Pipe

3⁄4" Adapters

To sprinklers

To sprinklers

Pipe from house water

To sprinklers

3⁄4" T

3⁄4" Elbow

3⁄4" T

are at least 6 inches (15.2 cm) higher than the highest point in the distribution system. (See "Control Manifold" Illustration on page 120.)

If you have underground utility lines, have your utility company mark line locations before laying out or digging at water distribution pipes and sprinkler head locations. Do not install sprinkler systems on city or other government-owned land without first getting permission.

With the control manifold in place, lay out and dig the trenches. Trenching is the hard part of installing a sprinkler system, so it's best to organize a volunteer work party of friends or relatives. If the soil is very dry and hard, run a sprinkler over the trench locations the night before you plan to dig.

The sprinkling will soften the soil so it is easier to dig. Use a square-end spade to cut the sod away before digging up the soil. Lay a plastic sheet next to the trench area and pile the sod and soil on the plastic. This will make refilling the trench easier and leave less mess on the lawn.

Also use a square-end spade to dig the trenches. The trenches should be V-

Pipe Trench

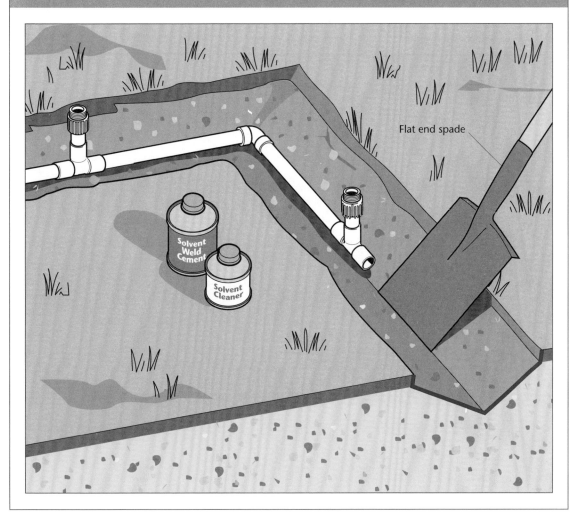

Flat end spade

Solvent Weld Cement

Solvent Cleaner

shaped and at least six inches (15.2 cm.) deep. Check the grade of the ditch, and slope the trenches away from the house if possible. Lay the pipes or flexible tubing in the trench and mark where the sprinkler heads and branch laterals will go. (See "Pipe Trench" Illustration above.)

If you must install a pipe under a sidewalk (on your property only: check first for permission if you are digging on city property), dig the trench on both sides of the walk. Then dig under the walk from both sides toward the middle. You may find it useful to use a small garden shovel, the type used when potting plants, to reach under the walk and complete the tunnel. Cover the ends of the pipe with duct tape to keep debris out while you push the pipe through the tunnel. Make the connections on both sides of the walk. (See "Trench Under Walk" Illustration on page 122.)

Use a plastic tubing cutter to cut the pipe at the sprinkler and lateral locations. When you have cut the pipes, install the riser Ts using the solvent weld or pipe clamp method. Be sure the riser Ts are secured so the riser will be perfectly vertical.

Trench Under Walk

Dig out the trench from both sides, then under the walk. Push pipe through hole under walk. To keep pipe debris-free, cover the ends of the pipe with duct tape.

At the control manifold use male threaded adapters to connect the distribution pipes to the manifold. Cut the risers so the tops are flush with the lawn. Screw the risers into the T connections, then install the sprinkler heads. When all connections are made on that zone, turn the water on and check for leaks. If there are no leaks proceed with the installation.

Now replace the soil in the trenches. Stand with one foot in the trench to hold the pipe down. Carefully replace the soil in the trenches, tamp it lightly and replace the sod.

To prevent freeze-up and damage to your sprinkler system in cold climates, you must drain or purge the system in the fall. If you neglect this step you may damage the pipes or sprinkler heads. The most effective way to purge the water from the system is to call in a pro who will use air pressure to blow out the lines. Using air pressure will pop up the sprinkler heads for complete water drainage.

CHAPTER 14

Finding a Repairperson

There will be a time when plumbing or other repairs lie beyond your abilities, in which case you'll require professional help. When is the best time to find a repairperson? Before you need one. Don't wait until you are standing in water to locate a good plumber. Shopping for repairpersons should be done when one has the time to check them out, not in the heat of an emergency situation.

Upon buying a house the homeowner should immediately seek the names of reliable companies he or she can call upon when an emergency, or a need for extensive repairs arises. Check with co-workers and neighbors to find local help. You will often find that a repairperson's reputation, good or bad, has become widely known in the community. People are eager to share both good and bad home repair experiences, and will share them with those who seek their advice.

Other referral sources for repairpersons include your local hardware store or lumberyard. These merchants deal with the pros in their area, and are often eager to assist their customers in finding help. Because a bad referral is bad for business, these merchants will usually not direct you to a shoddy worker or a rip-off artist who overcharges. In the United States you can also check with the Better Business Bureau to find if there are any complaints against the company in question.

Also, check credit references to find out if the contractor or repairperson pays his bills. In the U.S. a material supplier can file a lien against the homeowner's property for material purchases charged to that address. If major material expenses are involved, ask the repairperson to furnish lien waivers from suppliers when you pay the bill. A lien can be a troubling development when the supplier exercises his or her lien rights, or when the homeowner attempts to sell the property.

Get multiple bids for major projects. Don't make price the only criteria: a bid that comes in substantially below competing bids can be a red flag. Disreputable contractors, known in the trade as "low-ballers," purposely underbid a job to get the contract, then surprise the homeowner with inflated charges for "extras." Specify in the work contract that any changes in the original agreement be made only with a written agreement signed by both the contractor and the customer. Remember the old adage, "You don't always get what you pay for, but you never get more than you pay for."

Noted architect Frank Lloyd Wright advised that you should always ask repairpersons for references, and that you follow up by checking those references. The work speaks for the worker and Wright contended there is no other way to know the person except by his or her past work. Established repair people should have a long list of satisfied customers to whom they can refer new customers. Reliable repairpersons live on their referrals. Thoroughly check the references before you put the person's name and phone number in your file.

Glossary

A

Accessible—All plumbing fittings and valves must be accessible or reachable for service and repairs. Do not cover fittings or valves with ceiling tile, wallboard, etc., without providing an access panel. For example, the bathtub valve and shower assembly is usually accessed through an access panel on the backside of that wall, usually in a closet.

Adapter—A plumbing fitting used to connect two kinds or sizes of pipe.

Air gap—At a fixture such as a sink or tub, the distance between the faucet spout and the overflow or flood rim of the basin. The air gap prevents back siphoning of contaminated water into the water supply pipe. Also on dishwasher drainpipes.

Allen wrench—A hexagon-shaped tool used to tighten/loosen setscrews.

Angle-stop valve—A valve such as a fixture supply valve that causes water flow to make a 90° turn.

Anode—A rod installed in a water heater composed of one or more metals that protects the tank from corrosion.

Anti-siphon—Any device such as a valve that prevents back siphoning of contaminated water into the fresh water supply valve or pipe.

Approved—When plumbing has been inspected and accepted by the inspector, it has been approved.

B

Backflow—The reverse flow of water in a pipe.

Backwater valve—A valve in a sewer drain line that prevents fixtures from flooding because of sewer backup.

Ball cock—A toilet tank water valve opened or shut by a ball float.

Ball peen hammer—A hammer with a flat striking face on one side, and a rounded ball on the other. Used to flatten or "peen" rivets and shape metals.

Basin wrench—A long-handled wrench designed to reach past the sink basin to loosen retaining nuts or coupling nuts on the faucet tailpiece.

Branch—Any line or pipe other than the main, riser or stack.

Branch water service pipe—TK

Building drain—The lowest drainpipe in the house plumbing. It carries waste to the main sewer.

Burrs—Rough ridges left on the edges of pipe after cutting.

Bushing—A device inserted into a fitting that permits joining one pipe to a smaller pipe.

C

Cap—A fitting used to close off the end of a pipe.

Check valve—A valve that prevents water flow from reversing direction. Used in a sump pump discharge line to prevent water from flowing back into the sump pit.

Chlorinated polyvinyl chloride (CPVC) pipe—Rigid plastic pipe used for drain-waste-vent systems.

Chlorination—The addition of chlorine to water to kill bacteria and make the water safe for human consumption.

Circuit tester—A two-pronged electrical device inserted into the slots of a receptacle to test whether any current is present.

Cleanout plug—A removable plug in a drainpipe that permits access for cleaning the drain.

Closet auger—A flexible rod with a curved end that is inserted into a toilet's built-in trap and turned to remove clogs.

Cold water main—The primary pipe that conducts cold water to various fixtures in the home; typically ¾ inch.

Compression fitting—A pipe connection where a nut and a sleeve or ferrule is placed over a copper or plastic tube and is compressed tightly around the tube as the nut is tightened, forming a positive grip and seal without soldering.

Companion flange—One of two mating pipe fittings where the flange faces are in contact or separated only by a thin leak sealing gasket and are secured to one another by bolts or clamps.

Coupling—A fitting used to join two sections of pipe.

Cross Connection—When any freshwater pipe is connected to a drainpipe or other source of contamination.

D

Distribution box—The part of a septic system that takes the decomposed waste from the septic tank and feeds it into distribution pipes in the leach field.

Double T—A fitting used to join two branch pipes into a single pipe.

Drain auger—A steel cable with a flanged tip. The auger is inserted into a clogged drain and turned to clear away debris.

Drain—A pipe that carries sewage or wastewater away from the house.

Drop elbow—An elbow fitting that can be scraved to a well or framing member.

DWV—The drain-waste-vent system that collects plumbing waste and disposes it into the main sewer or septic system.

E

Elbow—An L-shaped fitting with two openings; also known as an L. Elbows come in various angles for various plumbing needs. Some of the more common elbows are:

Drop elbow—An ebloe fitting that can be screwed to a wall or framing member.

90' Drop L—TK

90' Long radius L—TK

90' Street L—TK

Emery paper—Paper or cloth coated with a fine abrasive used for cleaning or polishing pipe or metal.

Escutcheon—A cover placed over a pipe to conceal a hole in the wall.

F

Faucet—A valve that controls the water flow to a plumbing fixture.

Female adapter—A fitting that has inside pipe threads for connection with male pipe threads.

Female iron pipe (FIP)—Pipes and fittings with threads on the inside.

Filter—A device that removes contaminants and sediment from water, improving water purity and taste.

Fitting—Used to join two or more pipes; also used to change pipe size and/or direction.

Fixture drain—The drainpipe that connects a fixture trap with another drainpipe.

Fixture supply—Water supply pipes or risers that connect the fixture to the incoming water supply pipe.

Fixture supply tube—Tube that connects the fixture tailpiece to the shutoff valve.

Flapper—The moving part of a toilet's flush valve that seals the water into the tank or allows water to exit the tank for the flush cycle.

Flexible tubing—A braided stainless steel or PVC/polyester-reinforced hose connecting a faucet or toilet to the water supply stop valve. Serves as a riser but is much more flexible and easier to install.

Float ball—The floating ball connected to the ball cock inside a toilet tank that rises or falls with the water level, shutting off the ball cock as needed.

Float cup valve—The water valve that opens to flush the toilet. Replaces the old ball cock valve. Rather than a floating ball, the new valve uses a float cup that rises on the valve shaft to activate the valve.

Flood level rim—The top edge or rim of a tub or sink.

Flush bushing—For faucets, a valve connector device controlling water flow rate.

Flush valve—The valve inside the toilet tank that causes the toilet to flush when the flush lever is depressed.

Flux—A mixture of petroleum jelly and light acid, used to clean oxidation from metals prior to soldering.

G

Gas cock—A valve installed between the main gas line and a gas appliance.

Gate valve—A valve that allows full flow of water.

Globe valve—A gate valve with a curved chamber for adjusting the water rate.

Grab bar—A safety bar installed in a bathtub or shower to prevent falls.

Grade—The slope of a drainpipe toward the main drain, expressed as a fraction of an inch slope per running foot of drainpipe.

Ground Fault Circuit Interrupter (GFCI)—An electrical out-let with a device that senses any drop in current between the hot and neutral wires and shuts off the circuit to prevent electrical shock or fire.

H

Hacksaw—A fine-toothed saw designed to cut metal or plastic.

Hickey—An electrician's tool used to cut copper pipe.

Hose bib—An outdoor faucet, also used to supply washing machines.

Hot water main—The primary pipe that conducts hot water to various fixtures in the home; typically ¾ inch.

Hub (Shoulder)—The depth of a socket on a pipe fitting.

J

Joint compound—Paste material in a tube, applied to threaded joints in gas pipe.

K

Keyhole saw—A saw with a tapered blade designed to cut circular holes in wood, plaster or drywall.

L

L—see elbow.

Leach (seepage) field—An area of porous soil through which septic tank leach lines run, emptying treated waste.

Line stop—A valve installed in a water supply line to control water flow rate through the pipe.

M

Main—The drainpipe to which all other drains are con-nected.

Makeup—The length of pipe that extends into the hub of a fitting.

Male adapter—A fitting that has male pipe threads, used to connect to a fitting with female threads.

Male iron pipe (MIP)—Pipes and fittings with threads on the outside.

N

Nipple—A short length of pipe installed between couplings or other fittings.

No-hub—Cast iron pipe that has no connecting hub on one end. No-hub pipe is joined using sleeve-type connectors that are wrapped around the joint and secured with clamps tightened with a wrench.

O

O-Ring—A round rubber washer used to create a watertight seal, chiefly around valve stems.

P

Percolation test—A test done for septic fields to measure how effectively soil absorbs water.

Pilot—A small burner used to ignite the main burner of a gas appliance.

Pipe hangers—Steel, copper or plastic straps used to hold pipes in place for overhead assembly.

Pipe wrench—An adjustable wrench that has serrated jaws, designed to grip and hold round metal pipe.

Pliers—Tools designed to grip a fastener such as a bolt or nut, or various other workpieces. Types of pliers include:

Pliers, channel-type—Adjustable pliers with offset jaws designed to grip various plumbing fittings. Also referred to as water-type pliers

Pliers, needle-nose—Pliers with long, tapering jaws, designed to grip small objects or fasteners in narrow spaces.

Pliers, vise-grip—Locking pliers that have adjustable serrated jaws. An adjustment knob on the plier handle can be turned to grip a workpiece tightly until the locking mechanism is released.

Plumber's putty—A pliable putty used to seal joints between drain pieces and fixture surfaces.

Plumbing code—A set of rules governing the installation of plumbing systems. Developed by industry and government to ensure the safety and health of the public.

Plunger, flanged—A dome-shaped rubber cup with a wooden handle atop it used to unclog drains.

Polybutylene (PB) flexible piping—A flexible plastic tubing used in plumbing situations where rigid connections would be problematic.

Potable water—Water that is fit for human consumption.

Pressure regulator—A device installed at the water service entry to reduce the water supply pressure to a level acceptable for use within the house.

R

Reamer—A tool used to ream or smooth metal burrs left from cutting a pipe.

Reducer—A fitting used either to reduce or increase pipe size.

Relief valve drain—A pipe attached to the temperature and pressure pressure-relief (blowoff) valve to direct hot water or steam downward into a floor drain.

Retaining (grab) ring—A spring metal clip, shaped like a horseshoe, that is slipped into a groove on an axle, shaft or stem to secure two components together. For example, a retaining ring is used in a sleeve-type cartridge faucet.

Revent—A pipe installed specifically to vent a fixture trap. Connects with the vent system above the fixture.

Riser—A vertical pipe in the water supply system.

Riser tube—A flexible copper or plastic tube connecting a plumbing fixture to the water supply pipe.

Rough-in—Installation of the entire piping system and fixture supports prior to installation of the plumbing fixtures.

S

Septic tank—A watertight concrete, fiberglass or steel tank that breaks down the raw sewage from a house.

Service valve—A valve or faucet installed at a plumbing fixture. Used to shut off water supply while working on a fixture.

Shutoff valve—Any valve that shuts off the flow of water to a fixture, pipe or branch line.

Slip nuts—Connecting nuts used to join drain traps to pipes.

Socket wrench—A wrench is driven by a ratcheting handle that can be moved without removing the wrench from the nut. Sockets in sizes to fit various nuts can be snapped onto the ratchet handle.

Soil stack—The largest vertical drain line to which all branch waste lines connect; carries waste to the sewer line.

Solder—A metal alloy melted to join or repair metal surfaces; also, the act of melting solder into the joint.

Solvent weld—To join plastic pipe and fittings together with solvent-based cement.

Spring tubing bender—A coiled spring used to bend copper pipe. Slipped over the pipe, the spring prevents the walls of the pipe from collapsing during the bending process.

Stack vent—The portion of the main stack extending from the soil stack to the roof.

Stem washer—Used on a compression faucet, the washer is screwed to the end of the faucet stem to seal the joint between stem and seat.

Stop-and-waste valve—Valves with a cap on one side that can be removed to drain the pipes beyond the valve.

Street elbow—An elbow fitting with a female thread on one end and a male thread on the other.

Sump—A tank for collecting liquid wastewater below grade. A sump pump is used to pump the wastewater up to grade level, where it can flow away by gravity.

T

T—A T-shaped fitting with three openings that allows another pipe to be joined at a 90° angle, used to create branch lines.

T&P (temperature and pressure) relief valve—A safety valve installed on a water heater. If the heater malfunctions due to high temperature or internal pressure, the valve opens to prevent an explosive buildup. Also known as a blowoff valve.

Tailpiece—The short drainpipe located between the fixture and the trap.

Tank ball—Metal, rubber or plastic ball that fits into the flush valve of older toilet tanks to seal the tank.

Teflon Plumber's tape—Non-adhesive tape used to wrap pipe threads on water pipes to ensure a watertight seal.

Transition union—A fitting used to join plastic to metal water pipes.

Trap—A fitting in the drain system that provides a constant liquid seal, preventing sewer gases and odors from entering the house.

Tubing cutter—Tool used to cut metal pipe or plastic tubing.

U

Union—A fitting that joins two pipes but permits them to be disconnected later without cutting the pipe.

V

Vacuum—When the internal pressure of a pipe falls below atmospheric pressure.

Valve—A fitting that controls the flow of water.

Vent stack—Upper portion of the soil stack above the top-most fixture, through which gases and odors escape.

Vent draft hood—The hood atop a gas water heater that directs combustion gases upward via the vent pipe.

W

Waste pipe—Any drainpipe other than the toilet drain that carries away wastewater.

Water closet—Alternative name for the toilet.

Water hammer—Noise generated from hydraulic shock in a water pipe when the water flow stops abruptly, most common at a washing machine or dishwasher.

Water hammer muffler—A device installed near a fixture to absorb water hammer.

Water heater—Electric or gas-fired appliance that provides hot water for a house.

Water softener pipes—Pipes that connect the water service entry pipes to the water softener; typically ¾ inch.

Water service entry pipe—The main pipe carrying water into a house.

Wax ring—A ring of wax that joins and seals the toilet stool to the toilet drainpipe.

Y

Y—A Y-shaped fitting with three openings used to create branch lines.

Conversion Tables

Units of Length

in: inch; ft: foot; yd: yard   12 in = 1 ft; 36 in = 1 yd; 3 ft = 1 yd

Units of Weight

oz: ounce; lb: pound;   16 oz = 1 lb
g: gram; kg: kilogram   1,000 g = 1 kg

Units of Liquid Measure

oz: ounce; pt: pint; qt: quart   16 oz = 1 pt; 32 oz = 1 qt; 2 pt = 1 qt
gal: gallon   128 oz = 1 gal; 8 pt = 1 gal; 4 qt = 1 gal
mL: milliliter; L: liter   1,000 mL = 1L

CONVERSIONS

Multiply	By	To Get	Multiply	By	To Get

Length

Multiply	By	To Get	Multiply	By	To Get
in	25.4	mm	mm	0.039	in
in	2.54	cm	mm	0.003	ft
in	0.025	m	mm	0.001	yd
ft	304.8	mm	cm	0.394	in
ft	30.48	cm	cm	0.033	ft
ft	0.305	m	cm	0.011	yd
yd	914.4	mm	m	39.37	in
yd	91.44	cm	m	3.28	ft
yd	0.914	m	m	1.09	yd

Weight

Multiply	By	To Get	Multiply	By	To Get
oz	28.35	g	g	0.035	oz
oz	0.028	kg	g	0.002	lb
lb	435.6	g	kg	35.27	oz
lb	0.454	kg	kg	2.2	lb

Liquid Measure

Multiply	By	To Get	Multiply	By	To Get
oz	29.57	mL	mL	0.034	oz
oz	0.029	L	mL	0.002	pt
pt	473.2	mL	mL	0.001	qt
pt	0.473	L	mL	0.0003	gal
qt	946.4	mL	mL	33.8	oz
qt	0.946	L	L	2.11	pt
gal	3,785	mL	L	1.06	qt
gal	3.785	L	L	0.26	gal

Home Improvements Organizations

American Society of
Plumbing Engineers
8614 Catalpa Avenue,
Suite 1007
Chicago, IL 60656-1116
Phone: 773-693-ASPE (2773)
info@aspe.org

American Society of Sanitary
Engineers
901 Canterbury, Suite A
Westlake, OH 44145
Phone: 440-835-3040
Fax: 440-835-3488
info@asse-plumbing.org

American Standard
1 Centennial Ave.
Piscataway, NJ 08855
800-223-0068
800-387-0369 (Canada)
www.americanstandard-us.
com

American Waterworks
Association
6666 W. Quincy Ave.
Denver, CO 80235
Phone: 303-794-7711 or
800-926-7337
Fax: 303-347-0804
www.awwa.org

Bach Faucets
1110 Kamato Road, Unit 18
Mississauga, ON
L4W 2P3
www.bachfaucet.com

Canadian Tire Corp.
P.O. Box 770, Stn. K
Toronto, ON M4P 2V8
Phone: 1-800-387-8803
(English) or
1-800-565-3356 (French)
Fax: 1-800-452-0770
Feedback@canadiantire.ca

Crane
15 Crane Ave.
Stratford ON
N5A 6T3
Fax: 1-800-845-7158
www.crane.ca

Canadian Housing
Information Centre
700 Montreal Rd.
Ottawa, ON K1A 0P7
Phone: 1-800-668-2642 or
613-748-2367
Fax: 613-748-4069
chic@cmhc-schl.gc.ca

Color Caulk, Inc.
723 W. Mill St.
San Bernardino, CA 92410
Phone: 800-552-6225
Fax: 909-888-8497

Delta
55 East. 111th Street
Indianpolis, IN 46280
Phone: 317-848-1812
www.deltafaucet.com

Eljer Plumbingware, Inc.
14801 Quorum Dr.
Dallas, TX 75254
Phone: 972-560-2000 or
800-423-5537
www.eljer.com

Emco Corporation
620 Richmond St.
London, ON N6A 5J9
Phone: 519-645-3900
Fax: 519-645-2465
www.emcoltd.com

Kohler Plumbing Products
444 Highland Dr.
Kohler, WI 53044
Phone: 800-4-KOHLER
(800-456-4537) or
920-457-4441
www.kohler.com

Moen, Inc.
25300 Al Moen Dr.
P.O. Box 8022
North Olmsted, OH
44070-8022
Phone: 440-962-2000;
800-289-6636
Fax: 440-962-2770
www.moen.com

Moen Incoporated
2816 Bristol Circle
Oakville, ON L6H5S7
Phone: 905-829-3400
www.moencarded.ca

Plumbing Manufacturers
Association
1340 Remington Rd.,
Suite A,
Schaumburg, IL 60173
Phone: 847-884-9764
Fax: 847-884-9775
pmiadmin@pmihome.org

Index